C0-AOD-116

Land Stewardship through Watershed Management

Land Stewardship through Watershed Management

Perspectives for the 21st Century

Peter F. Ffolliott

University of Arizona
Tucson, Arizona

Malchus B. Baker, Jr., and
Carleton B. Edminster

USDA Forest Service
Flagstaff, Arizona

and

Madelyn C. Dillon and
Karen L. Mora

USDA Forest Service, CAT Publishing Arts
Fort Collins, Colorado

Kluwer Academic / Plenum Publishers
New York, Boston, Dordrecht, London, Moscow

ISBN 0-306-46698-8

233 Spring Street, New York, New York 10013

http://www.wkap.nl/

10 9 8 7 6 5 4 3 2 1

A C.I.P. record for this book is available from the Library of Congress

Printed in the United States of America

PREFACE

We must enhance the effectiveness of land stewardship and management of the world's natural resources to meet a growing global population's need for conservation, sustainable development, and use of land, water, and other natural resources. Ecosystem-based, multiple-use land stewardship is necessary when considering the present and future uses of land, water, and other natural resources on an operationally efficient scale. We need holistically planned and carefully implemented watershed management practices, projects, and programs to accommodate the increasing demand for commodities and amenities, clear water, open space, and uncluttered landscapes.

An international conference in Tucson, Arizona, from March 13 to 16, 2000, examined these needs and increased people's awareness of the contributions that ecosystem-based, multiple-use watershed management can make to future land stewardship. The conference was sponsored by the School of Renewable Natural Resources, University of Arizona; the College of Agriculture, University of Arizona; the Rocky Mountain Research Station, USDA Forest Service; the Research Center for Conservation of Water Resources and Disaster Prevention, National Chung-Hsing University, Taiwan; the Department of Forest Resources, University of Minnesota; the Center for Integrated Natural Resources and Agriculture Management, University of Minnesota; the Centro de Investigaciones Biologicas del Noreste, Mexico; the International Arid Lands Consortium; the USDA Natural Resources Conservation Service; the Bureau of Land Management of the Department of the Interior; the Salt River Project, Phoenix, Arizona; the Southern Arizona Chapter, Southwestern Section of the Society of American Foresters; and IUFRO Working Party 8.04.04, Erosion Control by Watershed Management.

At the conference, 35 speakers from research institutes, management agencies, and educational organizations in the United States and other countries presented invited synthesis papers in plenary sessions. More than 50 poster papers on watershed research projects, applied watershed management activities, and technology transfer mechanisms complemented the synthesis papers and broadened the conference scope. The proceedings of the conference, entitled "Land Stewardship in the 21st Century: The Contributions of Watershed Management," was published by the Rocky Mountain Research Station, USDA Forest Service.

This book expands on the general theme of the conference, integrates the published synthesis and poster papers, and provides supporting publications and reports. This more comprehensive information will now be available to a wider audience. The book's chapters focus on global watershed management perspectives, problems, and programs; a retrospective survey of watershed management, lessons learned, emerging tools and technologies, and locally led initiatives; the issues confronted when implementing a watershed management approach to land stewardship; the anticipated future contributions of watershed management to land stewardship; and the protocols necessary to realize the contributions of watershed management to land stewardship in practices, projects, and programs. As such, this book will be a valuable reference for researchers, managers, decision-makers, educators, students, and lay people with a keen interest in watershed management and improved future land stewardship. The book can also serve as a background to academic course work on the increasingly important topic of improved future land stewardship.

We gratefully acknowledge the valuable efforts made by the authors of the synthesis papers presented at the conference "Land Stewardship in the 21st Century: The Contributions of Watershed Management." Their reports summarized the state of the knowledge on a wide range of watershed management and land stewardship topics and established a benchmark for effectively considering the contributions that watershed management can make to improved future land stewardship.

Papers written by Kenneth N. Brooks, Karlyn Eckman, Carolyn Adams, Tom Noonan, Bruce Newton, Vicente L. Lopes, Hans M. Gregersen, and Allen L. Lundgren were the basis for the chapter on watershed management perspectives, problems, and programs. Elon S. Verry, James W. Hornbeck, Albert H. Todd, Wayne T. Swank, David R. Tilley, Robert L. Beschta, Luis A. Bojorquez Tapia, Exequiel Ezcurra, Marisa Mazari-Hiriart, Salomon Diaz, Paola Gomez, Georgina Alcantar, Daniela Megarejo, J. D. Cheng, H. K. Hsu, Way Jane Ho, T. C. Chen, Mohammed Shahbaz, B. Sunna, Satish Chandra, and K. K. S. Bhatia provided case studies that illustrated these perspectives, problems, and programs. We gratefully acknowledge Daniel G. Neary, Walter F. Megahan, James W. Hornbeck, Daniel P. Huebner, D. Phillip Guertin, Scott N. Miller, David C. Goodrich, and Michael D. Johnson for their contributions to the chapter on the retrospective viewpoint of watershed management. Authors of papers forming a core for the chapter on issues we will confront in the future include Roy Sidle, David B. Thorud, George W. Brown, Brian J. Boyle, Clare M. Ryan, Joe Gelt, and Eleanor S. Towns. Contributors of papers forming a basis for the chapter on contributions of watershed management to land stewardship were Michael Somerville, Dino DeSimone, J. E. de Steiguer, Warren P. Clary, Larry Schmidt, Leonard F. DeBano, and Jonathan W. Long. Bill McDonald, Hans M. Gregersen, William K. Easter, J. E. de Steiguer, Hanna Cortner, and Margaret M. Moote wrote key papers that were the basis for the chapter on future protocols.

We are also grateful to the authors of the poster papers presented at the international conference for supplementing and expanding on the synthesis papers. The poster papers reported on the results of watershed-related research projects, applied watershed management activities, and innovative technology transfer mechanisms for watershed-based information. We thank the many technical reviewers of the synthesis and poster papers for helping to ensure that the highest quality papers were published.

We extend appreciation to the Rocky Mountain Research Station for continued financial and collaborative support.

CONTENTS

1

INTRODUCTION

Three hundred years ago there were approximately 500 million people in the world. Today, the estimated human population is 6 billion—more than a tenfold increase. Even more striking is that global human population has doubled in the past 50 years. The impact on our world from this accelerating growth rate will determine how we manage the earth's ecosystems in the 21st century.

Every living thing requires sufficient space and resources to survive. Species with large numbers, such as humans, require extensive amounts of space and resources. When this happens, less space and resources are available for other living things, except those that benefit from human-modified habitats (Mathews, 1991; Postel, 1994; Salwasser, 1994). The effectiveness of land stewardship[1] must improve to meet our growing population's need for conservation, sustainable development, and use of land, water, and other natural resources.

We wrote this book to increase awareness about the many contributions that watershed management can make to improve future land stewardship. Along with a growing number of planners, managers, and decision makers, we believe that improved land stewardship is more likely to occur through a watershed management approach than through other strategic or tactical approaches. A watershed management approach toward land stewardship considers the environmental benefits associated with land, water, and other natural resource management and the products, services, and amenities demanded by society. Additionally, this approach identifies the relationship between environmental improvements and productivity increases over the long term.

1.1. WATERSHED MANAGEMENT APPROACH TO LAND STEWARDSHIP

Even experienced watershed managers have differing views about what watershed management means. In this book, we present a perspective of watershed management and a watershed management approach to land stewardship to help readers appreciate the contributions that this discipline can make to land stewardship. The perspective we have adopted is based on the following definitions and concepts (Brooks et al., 1992, 1994, 1997; National Research Council, 1999).

[1] The responsible and proper management of land, water, and other natural resources to enable their passage onto future generations in a healthy condition (Helms, 1998).

1.1.1. Watershed

A watershed is a topographically delineated area that is drained by a stream system—the total land area above some point on a stream or river that drains past that point. A watershed is also a hydrologic-response unit, a physical-biological unit, and a socioeconomic-political unit for management planning and implementation (Brooks and Ffolliott, 1995; Gelt, 2000; Thorud et al., 2000). A river basin is similar to a watershed, but it is larger. The Amazon River Basin, Mississippi River Basin, and Mekong River Basin include all of the lands that drain through these rivers and their tributaries into the ocean.

1.1.2. Watershed Management

Watershed management is the process of organizing and guiding land, water, and other natural resource use on a watershed to provide necessary goods and services, while mitigating the impact on the soil and water resources (National Research Council, 1999; Eckman et al., 2000; Gelt, 2000). Watershed management recognizes the interrelationships among soil, water, and land use, and the connection between upland and downstream areas.

Watershed-riparian relationships between stream-channel responses and impacts caused by natural or human-caused events on the surrounding watershed are part of watershed management. Watershed management involves socioeconomic, human-institutional, and biophysical interrelationships. Keeping these points in mind will help guide the design of watershed management practices and identify the institutional mechanisms needed to implement a watershed management approach to land stewardship.

1.1.3. Watershed Management Practices

Watershed management practices are changes in land use, vegetative cover, and other nonstructural and structural actions that are accomplished on a watershed to achieve ecosystem-based, multiple-use management objectives such as:

- Rehabilitating degraded lands;
- Protecting soil and water resources on lands managed to produce food, fiber, forage, and other products and to provide amenities such as the aesthetic landscape;
- Enhancing water quantity and quality; and
- Combinations of these objectives.

While many types of land-use practices occur on watersheds, when using an ecosystem-based, multiple-use oriented approach, resource production and environmental protection are equally important objectives (Brooks and Ffolliott, 1995). Assemblages of watershed management practices comprise watershed management projects, while groupings of watershed management projects are watershed management programs.

1.1.4. Watershed Management Approach

A watershed management approach to land stewardship incorporates soil and water conservation and land-use planning into a holistic, logical framework. This is accomplished by understanding the positive and negative impacts on people caused by the interactions of water with other resources. Also, it is important to understand that the nature and severity of these interactions are influenced by how people use the resources and the quantities of resources that they use. Water flows downhill regardless of how people might delineate their political boundaries. Therefore, the effects of these interactions follow watershed boundaries, not political boundaries. Activities on the uplands of one political unit—lands owned by an individual, family, community, or country—can significantly impact a downstream political unit.

Because these interactions cut across political boundaries, what one political unit considers sound use of natural resources, might be undesirable from a broader, societal point of view due to unacceptable downstream or off-site effects (Gregersen et al., 1987, 1996; Cortner and Moore, 2000). A watershed approach to land stewardship brings off-site effects into the analysis by considering the watershed boundaries. When off-site effects are present and the management costs and benefits are distributed among the political entities that accomplish and benefit from the watershed management practices, then ecologically sound management is also good economics.

A watershed management approach is designed to help resolve water, food, fiber, or other natural resource problems. Some of these problems and their causes, the structural and nonstructural solutions to these problems, and associated watershed management objectives are presented in Table 1.1. As indicated by Table 1.1, a watershed management approach provides a framework for conservation, sustainable development, and use of land, water, and other natural resources, while watershed management practices furnish the tools, technologies, and methods to make the framework operational. Various institutional mechanisms—regulations, market and non-market incentives, public and private investment—provide the means for implementing planned watershed management practices.

The logic and benefits of applying a watershed management approach to land stewardship has been well documented (Gregersen et al., 1987, 1996; Brooks et al., 1994, 1997; Eckman et al., 2000) and encompasses the multiple technical and socioeconomic dimensions that are necessary for effective land stewardship. Underlying this approach is an appreciation of conservation, sustainable development, and use of land, water, and other natural resources and an emphasis on decentralized and participatory approaches to land-use planning and management. Specifically, approaches that:

- Consider the interests of a wide range of stakeholders;
- Examine the benefits to better land stewardship by optimizing production and maintaining environmental integrity;
- Introduce more effective conflict resolution approaches from a sustainability perspective; and
- Recognize that future generations deserve to inherit landscapes that are capable of producing goods and services, while maintaining ecosystem health and sustainability.

A watershed management approach maintains a high sensitivity to the positive off-site effects associated with sustainable natural resource use, sufficient supplies of clean water, ecosystem protection from adequate instream water flows, access to biological and structural diversity, and carbon sequestration.

Table 1.1. Natural resource problems and their causes, structural and nonstructural solutions, and associated watershed management objectives.

Problems and Causes	Structural and Nonstructural Solutions	Associated Watershed Management Objectives
1. Inadequate water supplies caused by increasing population density, competing uses, pollution of water supplies, lack of access to supplies of water, inability to store water in periods of excess, and prolonged or frequent dry periods	a. Reservoir storage and water transport b. Water harvesting c. Reduce evapotranspiration d. Development of groundwater e. Cloud seeding	a. Minimize sediment delivery to reservoir site by maintaining vegetative cover b. Develop localized collection and storage facilities c. Convert from deep-rooted to shallow-rooted plant species or from conifers to deciduous trees d. Maintain high infiltration rates in groundwater recharge zones e. Maintain vegetative cover to minimize erosion
2. Flood damage caused by occupancy of flood-plain areas, increased surface runoff from watersheds, and inability to store and regulate stormflows	a. Reservoir storage b. Construct levees and improve channels c. Manage flood plain d. Revegetate disturbed areas	a. Minimize sediment delivery to reservoir site by maintaining vegetative cover b. Minimize sedimentation of downstream channels c. Zone lands to restrict human activities in flood-prone areas; minimize sedimentation of channels d. Implement reforestation or afforestation of denuded watersheds; encourage natural vegetation
3. Degraded watersheds caused by high rates of erosion and sedimentation	a. Erect erosion control structures b. Build contour terracing c. Revegetate	a. Maintain life of structures by revegetation and management b. Revegetate, stabilize slopes and terraces, and institute land-use guidelines c. Protect vegetative cover until site recovers; apply reseeding, fertilization, etc.
4. Polluted drinking water caused by improper development of wells, improper sewage treatment facilities, and contamination of surface water supplies	a. Develop alternative water supplies from wells, springs, etc. b. Treat water supplies	a. Protect groundwater from contamination b. Protect watersheds from contamination

Table 1.1. Continued

Problems and Causes	Structural and Nonstructural Solutions	Associated Watershed Management Objectives
5. Polluted streams and reduced fishery production caused by inappropriate land use, improper treatment of wastewater, and streambank degradation	a. Establish and the maintain vegetative cover on watershed b. Treat wastewater c. Protect streambank vegetation	a. Develop buffer strips along stream channels and manage to sustain vegetative cover on uplands b. Use natural systems (forests and wetlands) as secondary treatments c. Control livestock grazing and wildlife and develop management guidelines for riparian ecosystem
6. Food shortages caused by erosion and reduced capacity of the soil, population increases, waterlogging, and salinity resulting in losses of agricultural productivity	a. Develop agroforestry practices b. Increase cultivation c. Increase livestock production d. Import food from outside e. Drain waterlogged soil	a. Maintain site production; minimize erosion and nutrient losses; develop crops compatible with soils and climate of area b. Restructure steep hill slopes and other areas susceptible to erosion; use contour plowing, terraces, etc. c. Develop livestock herding-grazing systems for sustained production and productivity d. Develop forest resources for pulp and other wood products to provide economic base e. Maintain drainage ditches
7. Energy shortages caused by loss of fuelwood supplies because of reduced forest cover, population increases, loss of fossil fuels, inadequate infrastructure to distribute fuel, accumulation of sediment in reservoirs that provide hydropower and reduce the efficiency of turbines	a. Develop fuelwood resources b. Encourage agroforestry systems that include multipurpose and fast-growing trees c. Develop hydroelectric power projects such as mini-hydro projects, dams and reservoirs, etc.	a. Develop sustainable fuelwood plantations and guidelines that maintain productivity and protect soils b. Develop agroforestry practices that reduce erosion and protect fuelwood resources c. Minimize sediment delivery to stream channels and reservoir pools; apply vegetation and structural solutions

Source: Adapted from Brooks et al., (1990).

1.2. BOOK ORGANIZATION

The chapters in this book address:

- Watershed management perspectives, problems, and programs (Ch. 2);
- A retrospective viewpoint of watershed management to provide perspective on the earlier lessons learned for future consideration (Ch. 3);
- Future issues that planners, managers, and decision-makers will likely confront (Ch. 4); and
- Historical and proposed future watershed management contributions to land stewardship (Ch. 5).

Our concluding chapter discusses future protocols necessary to achieve better land stewardship through watershed management. We list references at the end of each chapter to provide readers with additional information sources that are applicable to that chapter. While not inclusive, these references will help planners, managers, and decision-makers learn more about the potential role of watershed management practices, projects, and programs in future land stewardship. A preview of the contents of the book's chapters follows.

1.2.1. Perspectives, Problems, Programs

By looking at the perspectives of watershed management in the United States and globally, we demonstrate how watershed management is effectively and efficiently used in conservation, sustainable development, and use of land, water, and other natural resources. Watershed management practices in Mexico, Taiwan, Jordan, and India are discussed to present a global perspective. Regional perspectives in the United States are from the Lake States, Northeast, and the East, the South and Southeast, the Pacific Northwest, and the Colorado River Basin. Issues of concern, past lessons learned, and future directions that might promote watershed management are reviewed. We believe that successful watershed management will advance only if the organizations that are responsible for its implementation as a land management strategy are highly adaptive, constantly seeking new information sources, and effectively using processes that foster innovation.

1.2.2. A Retrospective Viewpoint

It is important to acknowledge and understand past lessons to effectively accomplish current and future land stewardship through watershed management. The evolution of watershed management from ancient concepts written in Indian texts before 1,000 B.C. through contemporary theories published in late 20th century journals illustrates the broadening nature of watershed management. Research findings and management experiences demonstrate the effects of vegetative management on streamflow regimes, erosion and sedimentation processes, water-quality constituents, and other natural resources. The resulting body of knowledge on hydrologic, nutrient, and energy cycles has been incorporated into land, water, and other natural resource management efforts using best management practices[2].

Implementation of watershed management practices reduces or prevents ecosystem pollution and other environmental problems. Advances in computer modeling and simulation

[2] A practice or a combination of practices that are determined by a state or a designated planning agency to be the most effective and practicable means (including technological, economic, and institutional considerations) of controlling point and nonpoint source pollution at levels compatible with environmental quality goals.

techniques help to evaluate the cumulative effects of these practices. We discuss other emerging tools and technologies that facilitate the capture, storage, and use of spatial data sets for improving the scientific understanding of watershed processes. Additionally, we examine the importance of sociocultural perspectives regarding the development of watershed management partnerships.

1.2.3. Issues to be Confronted in the Future

This chapter considers future global and regional United States watershed management issues that are likely to be confronted by people and societies. Many of these issues center on the current and future status of the watershed management planning process. This highly interactive planning process is complex and often difficult because of the physical, biological, and social interactions that are the foundation of watershed management (Brooks et al., 1992, 1994, 1997; Eckman et al., 2000). People have a responsibility to collaborate to conserve natural resources and, when necessary, preserve their integrity for future generations. The outcome of this joint planning effort is natural resource sustainability.

1.2.4. Watershed Management Contributions to Land Stewardship

Chapter 5 presents the core concepts of the book. Effective watershed management practices contribute to land stewardship by sustaining the physical and economic flows of crucial natural resources. These natural resources include tangible commodities—water, timber, and forage for livestock production—and intangible values—scenic quality, experiencing solitude, and existence and bequest values. Securing clean water will continue to be a significant watershed management contribution to land stewardship, as will maintaining the future health and stability of sensitive watershed-riparian ecosystems by recognizing the complex ecosystem responses encompassed. The contribution of watershed management to future land stewardship must reverse the adverse impacts on fragile riparian sites caused by the actions of people on surrounding watersheds, while enhancing the opportunities for investment, employment, and income-generation.

1.2.5. Future Protocols

To achieve better land stewardship, future protocols will likely focus on responding to increased demand for land, water, and other natural resources, anticipating future watershed conditions and maintaining landscape integrity, and implementing effective policies. Water will likely unify many integrated watershed management elements. However, effective policies that incorporate ecological understanding into their structure and promote democratic ideals will also be necessary. Guidelines to achieve this include immediate integration of the political process, building bridges to citizens, examining laws, rights, and responsibilities, strengthening administrative capacity, and looking beyond the watershed to a broader scale.

1.3. SUMMARY

The human population has increased more than tenfold in the past 300 years and has doubled in the past 50 years. Ramifications of our accelerating growth rate will influence how the earth's ecosystems will be managed in the 21st century. The effectiveness of land stewardship must improve to meet the growing need for conservation, sustainable development, and use of land, water, and other natural resources. Increasing people's awareness of the contributions that watershed management can make to future land stewardship through an exploration and evaluation of potential is the purpose of this book. Better land stewardship will more likely be achieved through a watershed management approach than through other strategies because this approach allows for a balance between the environmental benefits associated with land, water, and other natural resource management and the products, services, and amenities demanded by society. The watershed management approach also helps to identify key relationships between environmental improvements and productivity increases in the long-term.

REFERENCES

Brooks, K. N., Ffolliott, P. F. "Watersheds as Management Units: An International Perspective." In *Rangelands in a Sustainable Biosphere: Proceedings of the Fifth International Rangeland Congress*, N. E. West, ed. Denver, CO: Society for Range Management, 1995, pp. 167-169.

Brooks, K. N., Ffolliott, P. F., Gregersen, H. M., DeBano, L. F., *Hydrology and the Management of Watersheds*. Ames, IA: Iowa State University Press, 1997.

Brooks, K. N., Ffolliott, P. F., Gregersen, H. M., and Easter, K. W., *Policies for Sustainable Development: The Role of Watershed Management*. Washington, DC: U.S. Department of State, EPAT Policy Brief 6, 1994.

Brooks, K. N., Gregersen, H. M., Ffolliott, P. F., Tejwani, K. G. "Watershed Management: A Key to Sustainability." In *Managing the World's Forests: Looking for Balance Between Conservation and Development*, N. P. Sharma, ed. Dubuque, IA: Kendall/Hunt Publishing Company, 1992, pp. 455-487.

Brooks, K. N., Gregersen, H. M., Lundgren, A. L., Quinn, R. M., *Manual on Watershed Management Project Planning, Monitoring and Evaluation*. College, Laguna, Philippines: ASEAN-US Watershed Project, 1990.

Cortner, H. J., Moote, M. A. "Ensuring the Common for the Goose: Implementing Effective Watershed Policies." In *Land Stewardship in the 21st Century: The Contributions of Watershed Management*, P. F. Ffolliott, M. B. Baker, Jr., C. B. Edminster, M. C. Dillon, and K. L. Mora, tech. coords. Fort Collins, CO: Rocky Mountain Research Station, USDA Forest Service, Proceedings RMRS-P-13, 2000, pp. 247-256.

Eckman, K., Gregersen, H. M., Lundgren, A. L. "Watershed Management and Sustainable Development: Lessons Learned and Future Directions." In *Land Stewardship in the 21st Century: The Contributions of Watershed Management*, P. F. Ffolliott, M. B. Baker, Jr., C. B. Edminster, M. C. Dillon, and K. L. Mora, tech. coords. Fort Collins, CO: Rocky Mountain Research Station, USDA Forest Service, Proceedings RMRS-P-13, 2000, pp. 37-43.

Gelt, J. "Watershed Management: A Concept Evolving to Meet Needs." In *Land Stewardship in the 21st Century: The Contributions of Watershed Management*, P. F. Ffolliott, M. B. Baker, Jr., C. B. Edminster, M. C. Dillon, and K. L. Mora, tech. coords. Fort Collins, CO: Rocky Mountain Research Station, USDA Forest Service, Proceedings RMRS-P-13, 2000, pp. 65-73.

Gregersen, H. M., Brooks, K. N., Dixon, J. A., Hamilton, L. S., *Guidelines for Economic Appraisal of Watershed Management Projects*. Rome: Food and Agriculture Organization of the United Nations, FAO Conservation Guide 16, 1987.

Gregersen, H. M., Brooks, K. N., Ffolliott, P. F., Henzell, T., Kassam, A., Tejwani, K. G., Watershed management as a unifying framework for researching land and water conservation issues. Land Husbandry 1996; 2:23-32.

Helms, J. A., ed., *The Dictionary of Forestry*. Bethesda, MD: Society of American Foresters, 1998.

Mathews, J. T. ed., *Preserving the Global Environment: The Challenge of Shared Leadership*. New York: W. W. Norton and Company, 1991.

National Research Council, *New Strategies for America's Watersheds*. Washington, DC: National Research Council, National Academy of Sciences, 1999.

Postel, S. "Carrying Capacity: Earth's Bottom Line." In *The State of the World 1994: A Worldwatch Institute Report on Progress Toward a Sustainable Society*, L. Starke, ed. New York: W. W. Norton & Co., 1994.

Salwasser, H. *Ecosystem management: Can it sustain diversity and productivity.* Journal of Forestry 1994; 92(8): 6-10.

Thorud, D. B., Brown, G. W., Boyle, B. J., Ryan, C. M. "Watershed Management in the United States in the 21st Century." In *Land Stewardship in the 21st Century: The Contributions of Watershed Management*, P. F. Ffolliott, M. B. Baker, Jr., C. B. Edminster, M. C. Dillon, and K. L. Mora, tech. coords. Fort Collins, CO: Rocky Mountain Research Station, USDA Forest Service, Proceedings RMRS-P-13, 2000, pp. 57-64.

2

PERSPECTIVES, PROBLEMS, PROGRAMS

In this chapter, we consider how watershed management practices, projects, and programs are effectively and efficiently incorporated into the conservation, sustainable development, and use of land, water, and other natural resources. We present a global view of watershed management and regional views of watershed management in the United States. Case studies from Mexico, Taiwan, Jordan, and India are examples of how watershed management is practiced on a global-scale. Regional case studies in the Lake States and Northeast, the Southeast, the Pacific Northwest, and the Colorado River Basin illustrate watershed management perspectives, problems, and programs in the United States. Future directions to follow to promote better land stewardship through watershed management are presented in these studies.

2.1. GLOBAL VIEWS

We view watersheds globally as useful systems for planning and implementing natural-resource and agricultural development programs. In many countries, watershed practices focusing on soil and water conservation and watershed rehabilitation originated during their colonial periods (Brooks and Eckman, 2000; Neary, 2000). Following independence, many newly formed central and regional governments promoted large-scale hydropower and other water-resource development projects to facilitate rapid progress toward independent economic development.

By the 1950s and continuing through the 1980s, watershed management programs in many developing countries centered on restoring land and water systems that the increasing demand for natural resources had degraded. Additionally, it was important to protect earlier investments made in water-resource development. Many of these programs were narrowly focused, temporary solutions to degradation problems, with little thought to the spatial and temporal aspects of watershed management. Recently, many developing countries have incorporated participatory methods of planning and interdisciplinary approaches to management into their watershed programs to achieve a sustainable approach to overcoming degradation problems.

National governments, bilateral and multinational agencies, and nongovernment organizations have initiated various watershed practices, projects, and programs to achieve sustainable flows of natural resources and agricultural benefits from watershed lands. However, the benefits of these interventions do not depend solely on the physical and biological characteristics of the watersheds. The cultural background of rural populations inhabiting the watersheds; the nature of the governments; and they must effectively integrate other institution, economic, and social factors to meet environmental, economic, and social goals and objectives (Brooks et al., 1992, 1994, 1997). We show how technical and sociocultural factors are globally interrelated by examining the watershed management perspectives, problems, and programs in the Basin of Mexico, the mountainous watersheds of Taiwan, the Azraq Basin of Jordan, and the Indian Subcontinent. These case studies represent diverse global views of watershed management.

2.1.1. Basin of Mexico

The history of the Basin of Mexico, which humans have occupied for almost 4,000 years and is the present site of Mexico City, is one of growth, collapse, and rebirth. These cycles are rooted in the depletion of local water supplies and other natural resources, and the resulting dependence on imported resources. Mexico City and its neighboring municipalities—one of the largest megalopolises in the world—dominate the basin. Because of its large population, and their demand for land, water, and other natural resources, many people inside and outside the Basin of Mexico have questioned the viability of this area (Bojorquez et al., 2000). Available and reliable groundwater reserves, the primary source of water in the basin, are particularly limiting. To avoid an environmental crisis in this area, a multi-basin approach to watershed management is necessary to integrate a water-resource management strategy with the sustainable use of other natural resources (Aguilar et al., 1995; Ezcurra and Mazari-Hiriart, 1996).

2.1.1.1. The Setting

The Basin of Mexico covers 7,500 km^2 at the southern end of the Mexican Meseta Central (central plateau) and encompasses the Federal District of Mexico and portions of the states of Mexico, Hidalgo, Tlaxcala, and Puebla. Annual, mostly monsoonal, precipitation in the mountains surrounding the basin varies from a high of 1,200 mm to a low of 800 mm, depending on elevation—the valley floor receives much less precipitation. About 50% of the precipitation percolates to the basin's groundwater aquifers (Bojorquez et al., 2000).

The surrounding mountains, the highest of which are above 2,700 m, support forests and woodlands of intermingling fir, pine, and juniper. More than 25% of the people in Mexico are concentrated in the Basin of Mexico. While the average growth rate of the megalopolis was greater than 5% from the early 1950s through 1980, it has been decreasing in recent years (Ezcurra and Mazari-Hiriart, 1996). Widespread land-cover changes are rapidly occurring in the basin. Urban growth, deforestation, agriculture, and ranching interests are continually competing for the available land, water, and other natural resources, with resulting environmental conflicts.

2.1.1.2. Major Watershed Management Problems

Insufficient groundwater reserves, land subsidence caused by over-pumping groundwater, groundwater pollution, and wastewater contamination of the limited surface water are the major watershed management problems in the Basin of Mexico (Ezcurra and Mazari-Hiriart,

1996; Bojorquez et al., 2000). Interestingly, until the middle 1960s, groundwater extractions provided enough water to the people of Mexico City and vicinity. However, due to the demands of a growing population, the groundwater-extraction rate is now higher than the natural recharge rate of the main aquifers. Consequently, the people in this region rely on surface water collected within the basin and water transfers from outside the basin to augment the water supply.

Stormflow runoff, industrial wastewater, and sewage for about 75% of the people are transported from the basin via sewers, open canals, and a deep drainage system. Because the remaining population disposes their often untreated sewage through septic tanks and absorption wells (Bojorquez et al., 2000), many contaminants are released at the surface and migrate or are carried down through the soil by percolating water to the groundwater aquifers. Compounding the problem of contamination is the limited capacity of the treatment plants in the basin. The total amount of wastewater treated in these plants is equivalent to less than 10% of the water used. Other watershed management problems in the basin are the siltation of the drainage system and widespread flooding due to deforestation on the surrounding mountains (Aguilar et al., 1995; Ezcurra and Mazari-Hiriart, 1996). Unfortunately, the current infrastructure of the megalopolis is insufficient to handle these problems in most situations.

2.1.1.3. Watershed Management Programs

Suelo de Conservacion Ecologica, established by a presidential decree in 1930, is an area in the Federal District of Mexico (Bojorquez et al., 2000). Known locally as the "conservation land," *Suelo de Conservacion Ecologica* covers almost 900 km^2, which is about 60% of the Federal District, mostly in the mountain ranges to the south and southwest of Mexico City. The area encompasses 25 designated watersheds, the main groundwater-recharge areas in the basin, and 500 km^2 of forests, shrublands, and grasslands. The Natural Resources Commission of the Federal District is the government agency responsible for managing the *Suelo de Conservacion Ecologica*. This commission is charged with:

- Protecting the natural vegetation, ecosystem habitats, and groundwater-recharge areas in the district;
- Delineating land-use patterns that maximize consensus and minimize the environmental conflicts surrounding land use; and
- Formulating policies for the conservation, sustainable development, and use of land, water, and other natural resources in the *Suelo de Conservacion Ecologica*.

The commission must discharge their responsibilities within the complexity of widespread, often unplanned urban development and frequent overgrazing by livestock, illegal tree harvesting, and uncontrolled recreation activities.

The biggest challenges facing the *Suelo de Conservacion Ecologica* are ensuring adequate water supplies and wastewater disposal. Watersheds with the highest water surplus are also the lands most threatened by urbanization. Therefore, a multi-basin approach to water resource management is necessary to meet water-supply challenges in the immediate future (Ezcurra and Mazari-Hiriart, 1996; Bojorquez et al., 2000). By 2015, transportation of water from outside the basin will likely be needed to satisfy anticipated water demands. The availability of external water resources will hopefully lower the current rate of groundwater extraction in the basin, thereby, reducing the subsidence problems in the megalopolis. The commission has also formulated plans to improve wastewater disposal. Such provisions include protecting and sustaining other natural resources in the Basin of Mexico, and its neighboring basins, using an integrated watershed management approach. However, securing the capital necessary to implement these plans in the Basin of Mexico remains a problem.

2.1.2. Mountainous Watersheds of Taiwan

Few places in the world experience watershed management problems similar to those that confront the people of Taiwan. Heavy torrential rains in the typhoon season, steep mountainous topography, young and unstable geologic formations, highly erodible soils, frequent earthquakes, and improper land-use practices contribute to widespread landslides, debris flows, and floods in Taiwan (Cheng et al., 2000). With growing public and government support, the country's steadily expanding watershed management programs have helped to alleviate the devastating impacts of these disasters, while enhancing the sustainable use and management of land, water, and other natural resources (Figure 2.1). However, the accelerating economic growth occurring in downstream urban and industrial areas will continue to increase the demand for high-quality water. The challenges to watershed management in Taiwan are to mitigate disasters, sustain economic development, and integrate watershed management programs and other government services to meet the needs and expectations of its people.

2.1.2.1. The Setting

Rugged mountains and hills cover two-thirds of Taiwan, with steep and geologically unstable slopes that typically exceed 45%. Average annual rainfall is 2,500 mm, which is almost 3 times more than the global average—more than 3,000 mm of precipitation fall in the high mountain regions. About 80% of the rainfall is concentrated in the May-to-October

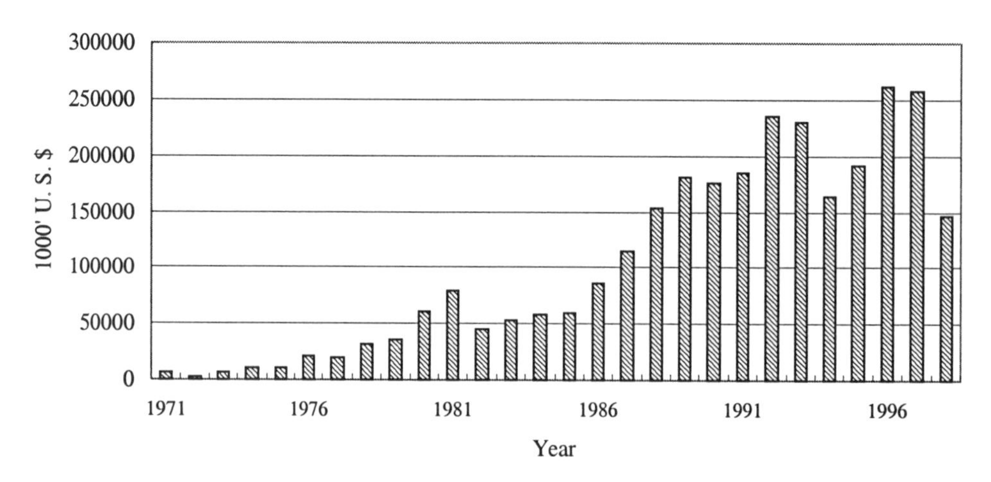

Figure 2.1. Government expenditures in Taiwan's watershed management programs from 1971 to 1998 (from Cheng et al., 2000).

typhoon season, when as much as 1,000 mm can fall in a single day (Cheng et al., 2000). More than 1,985 mm of rainfall was measured at one high-elevation weather station in a 48-hour period during Typhoon Herb in 1986. Unfortunately, because of the steep topography and limited river-system lengths, only 20% of the rainfall is captured for use by the country's large population—most of it flows quickly to the sea (Tung and Hong, 2000). Additionally, landslides are often caused by high quantities of rain falling on the unstable geological formations.

Dense montane forests—conifers, hardwoods, and bamboos—cover nearly 60% of Taiwan. One-fourth of the country's surface is devoted to upland, paddy, and other farming practices. One-half of the farming practices are accomplished on slopes steeper than 30%—much of which is subject to high erosion rates when exposed. The remainder of the country is grasslands, urban developments, industrial activities, and inland water surfaces.

2.1.2.2. Major Watershed Management Problems

Even with high rainfall amounts, the availability of usable water in Taiwan is not well distributed within a year. During the typhoon season, excess water is saved in reservoirs for use in the drier seasons. However, rapid economic growth is increasing the demand for water, existing reservoir and water-delivery systems are rapidly becoming overtaxed, and potential new reservoir sites are almost nonexistent (Tung and Hong, 2000). Compromises will be necessary to continue sound economic growth, while environmentally managing the country's water resources. Landslides and the degradation of water quality are the major watershed management problems in Taiwan (Cheng et al., 2000). Landslides contribute large amounts of sediment and other debris to stream channels and then into river systems. Depending on where these materials are deposited:

- Streambeds rise;
- Reservoirs and irrigation system capacities are reduced and hydropower facilities are damaged; and
- Drinking water supplies are at risk.

A major goal of watershed management in Taiwan is to mitigate the heavy siltation in streams and reservoirs to improve the flow of high-quality water to users. Heavy application of fertilizers and pesticides on hillslope agricultural lands also contribute to water degradation and, therefore, is addressed in the country's watershed management program.

2.1.2.3. Watershed Management Programs

Forest clearing and road construction followed by improper land use accelerates the occurrence of landslides, debris flows, and downstream problems. The government in Taiwan has been implementing watershed management programs throughout the country over the past 50 years to combat the damaging effects of these actions (Cheng et al., 2000). Key components of these programs are:

- Implementing sustainable and environmentally sound forest management practices, terracing, and other soil conservation methods on steep cultivated hillslopes;
- Regulating construction and maintenance of roads to achieve and retain their stability;

- Continuing landslide-prevention measures and rehabilitating impacted sites; and
- Stabilizing stream channels.

The backbone of Taiwan's watershed management programs is the effective regulation of land-use practices on fragile landscapes. Legislation, policy, and institutional arrangements exist to implement effectively water-conservation and watershed management programs (Cheng et al., 2000).

Landslide prevention, rehabilitation of impacted sites, and stream-channel stabilization receive for the largest proportion of the country's watershed management budget (Cheng et al., 2000). Landslide prevention measures include:

- Excavating unstable materials from landslide-prone sites;
- Accomplishing proper drainage of overland flow on landslide-prone sites; and
- Constructing structures such as retaining walls, buttresses, and pilings.

Installing engineering structures followed by site-specific revegetation rehabilitates impacted sites. Constructing check dams, submerged sills, bank-protection dikes, and stream regulation stabilizes stream-channels. However, despite these efforts, in the mountainous watersheds of Taiwan, people continue to experience landslides, debris flows, and floods. Consequently, watershed managers, hydrologists, and decision-makers continue to seek innovative ways to plan and implement effective watershed management programs to safeguard the lives of Taiwan's people. Ultimately, affecting behavior through education is necessary to generate and sustain the public's appreciation and support of and participation in watershed management programs in Taiwan.

2.1.3. Azraq Basin of Jordan

Water scarcity—historically restricting economic development in the Azraq Basin in eastern Jordan—will likely be a limiting factor until the region's water resources are adequately evaluated, actively managed, and equitably allocated. Therefore, an evaluation of existing water resources and the development of additional water supplies is the goal of the integrated watershed management programs in the Badia[3] region of eastern Jordan (Dutton et al., 1998; Shahbaz and Sunna, 2000). Covering about 80% of Jordan's total area, the Badia is a sparsely populated region of traditionally mobile pastoralists, who move their herds on the harsh terrain to follow either the seasons or localized rainfall. Most watershed management programs in the Badia region are centered in the Azraq Basin.

2.1.3.1. The Setting

The Azraq Basin, about 12,750 km^2 in area, is in the northeastern Badia region. The town of Azraq—in the center of the basin—was built around an oasis formed by the emergence of a groundwater spring (Dutton et al., 1998; Shahbaz and Sunna, 2000). Groundwater that discharges at this oasis is a major source of water for northern and eastern Jordan. The land surface of the basin is composed of gently undulating low hills on which only few herbaceous plants grow. Wind and water erosion washes most of the weathered basaltic soils into low basins. The climate of the region is continental, with large seasonal and diurnal temperature fluctuations. Average annual precipitation ranges from more than 250 mm along the Syrian border in the north to less than 50 mm in the south.

[3] An Arabic word that describes dryland regions where the annual rainfall averages less than 200 mm.

Over recent decades, increases in the region's human and animal populations, expansion of irrigated agriculture, and the desire of the inhabitants in nearby cities and towns—including Amman, the country's capital—for a higher standard of living has caused increasing groundwater extraction from the basin. Increased groundwater extraction has resulted in a depression in the regional groundwater table, and oases degradation in Azraq and the surrounding areas. Although intermittent, short-duration surface-water flows exist in the winter, most of the water flows onto mudflats and evaporates, thus, this not a dependable water supply for the region's inhabitants.

2.1.3.2. Major Watershed Management Problems

The overriding watershed management problem in the entire Badia region, including the Azraq Basin, is the need to sustain and replenish the groundwater reserves. The hydrologic relationship between the limited surface water and the more abundant groundwater resources is unclear. However, considering current and projected overdrafts, it is certain that the quantity of water percolating from the surface to the water table is insufficient to sustain the groundwater resources. Other watershed problems in the Azraq Basin (Dutton et al., 1998) include:

- Locating areas for further groundwater development;
- Identifying sites for exploring mineral, natural gas, and oil deposits; and
- Delineating areas for agricultural expansion.

To sustain the people's use of the limited natural resources in the basin and maintain the integrity of the fragile landscapes, these activities must be accomplished without environmental degradation. This is the focus of the region's watershed management programs.

2.1.3.3. Watershed Management Programs

To evaluate the potential of water and the other natural resources in the Azraq Basin, a watershed management approach to land stewardship is being implemented to satisfy the increasing demand for these resources (Dutton et al., 1998; Shahbaz and Sunna, 2000). To meet this objective, the government of Jordan has initiated a series of sequential studies. Specialists in water resources, geology, geophysics, mineralogy, agriculture, and environmental quality are collaborating to integrate and analyze the information available on land forms, drainage patterns, geomorphology, and soils in the basin. One result of this work has been the preparation of updated geological maps delineating the location of major faults that directly influence groundwater movement.

Development of a three-dimensional geological model to facilitate interpretation of information on the drainage network and other hydrological features of the Azraq Basin is another noteworthy outcome of the program. The continuing monitoring phase is:

- Identifying the changes in the physical, chemical, and biological properties of the surface and groundwater resources in relation to increasing development;
- Determining the long-term impacts of natural ecological processes and human interventions on these resources; and
- Defining measures to adopt to prevent groundwater depletion and pollution.

As the watershed management programs continue, a Geographic Information System (GIS) is being developed to store, manipulate, and display the baseline information and data sets collected in the Azraq Basin. GIS use will encourage cooperation and the necessary communication between the many agencies and organizations involved within the programs (Shahbaz and Sunna, 2000). Overall success of the watershed management programs is measured largely by the efficiency achieved in managing the land, water, and other natural resources, within a comprehensive and responsible framework, to meet the needs of the local inhabitants and the people of Jordan. The long-term goal of the integrated watershed management programs in the Azraq basin and, more generally, the Badia region is the extrapolation of these programs to the other basins in the country.

2.1.4. India

With a rapidly growing human population and improving living standards, the food requirements of India's people must increase by 50% over the next 20 years. However, expansion of the agricultural sector to meet these demands can be only achieved by rehabilitating watershed lands degraded by excessive soil erosion. Disturbance of the protective vegetative cover is caused by agricultural cultivation, livestock grazing, and burning to produce higher productivity levels (Chandra and Bhatia, 2000; Gowda and Jayaramaiah, 2000). Considering the increasing human population and accelerating agricultural and industrial development, more efficiently using the country's water resources is also necessary. Restoring the degraded watersheds and improving the availability and delivery of water resources is related to the country's sustainable land, water, and other natural resource management in the country. Effective watershed management is paramount to meet this challenge.

2.1.4.1. The Setting

India is the second most populous and seventh largest country—in total area—in the world. Bounded by the snow-capped Himalayas in the north, the country stretches southward and, at the Tropic of Cancer, tapers off into the Indian Ocean between the Bay of Bengal to the east and the Arabian Sea to the west. India's population is estimated at 1 billion people, and it is expected to increase to 1.1 billion people over the next 10 years (Chandra and Bhatia, 2000). It is not surprising therefore, that the country's limited land base (3.3×10^6 km^2) is subject to ceaseless overexploitation resulting in large-scale soil losses and other types of environmental degradation. Additionally, from this fixed land-base, large tracts of forests and grasslands have been cleared or degraded.

Precipitation is highly variable in time and space throughout the country. Annual rainfall varies from 11,000 mm at Cheerapouji, Meghalaya, to almost 100 mm in western Rajasthan. Not all of the major river systems are gauged to measure the volume of their water flows. Nevertheless, the amount of water flowing in India's rivers, which is available for beneficial use, is much less than the total water flows is the rivers. Unfortunately, because of limitations imposed by physiography, interstate issues, and the lack of appropriate technologies to harness the water resources economically, increasing these water flows is unlikely.

2.1.4.2. Major Watershed Management Problems

Many of the watershed management problems confronted by other countries are also issues in India. However, the Subcontinent faces most of these problems on a continuing and often simultaneous basis. India is not achieving efficient allocation, distribution, and use of the available water on a sustainable and countrywide basis using adequate environmental safeguards (Chandra and Bhatia, 2000). Groundwater quality has also been declining in many parts of the country and, therefore, is a concern and a threat to the continued availability of high-quality water for human consumption (Kumar, 1992). Additionally, management of water resources by different state governments is not often based on comprehensive river-basin planning and surface and groundwater resources are not collaboratively managed to ensure optimal allocations. Development projects for surface and groundwater resources continue to be planned independently (Mishta and Singh, 1992).

Droughts are of particular concern to watershed managers in India (Agarwal, 1992; Chandra, 1993; Jha et al., 1993; Chandra and Bhatia, 2000). Many drought-prone areas in the country are also the most heavily populated, therefore, drought-conditions cause considerable human suffering and economic loss. Moving emergency food, fodder, water, and other relief measures to the affected drought areas is extremely costly. Indirect losses from crops not planted, lands abandoned, and land-use changes also cause severe economic burdens. The emergency financial assistance from the state or central government to those affected by droughts impacts the country's overall economy. Conversely, the devastating floods associated with the summer monsoon are also a problem of major consequence in India. Unfortunately, flood-control measures are not an integral part of the country's water-resources management (Haque, 1992).

Spreading urbanization, much of which is unplanned, is occurring at a steadily increasing rate throughout the Indian Subcontinent. The result of this urban sprawl is the disappearance of productive watershed lands. Urban development is also displacing agricultural lands, grasslands for livestock grazing, recreational opportunities, wildlife habitats, and open spaces (Bhatia, 1992; Chandra and Bhatia, 2000). Often the agricultural lands and natural resources are not considered in the water-resource management planning efforts typically followed in India or, if considered, they are inadequately addressed. A comprehensive planning process is needed to holistically continue the country's urban development.

Another major watershed management problem in India is sociocultural in nature. Activities in watershed management programs for the country's women do not consider them equal to their male counterparts (Chandra and Bhatia, 2000). Women are disadvantaged because their contributions to the rural economy are generally unappreciated. Consequently, they do not receive equitable compensation in personal wages or in benefits accrued through commodity production. Policymakers have recognized the importance of increasing women's participation in watershed management programs and, as a result, efforts are being made to elevate the status of women to foster equitable partnerships.

2.1.4.3. Watershed Management Programs

Since India's independence in 1947, watershed management, hydrology, and natural resource management has improved. Advancements, successful in alleviating many of the country's acute watershed management problems include:

- Establishing hydro-meteorological networks on large river basins by the central and the state governments;
- Implementing nuclear and remote-sensing techniques, data loggers, and micro-processors to collect baseline data at remote sites;
- Using high-speed computers to develop real-time solutions to complex and often dynamic watershed-management problems;

- Collecting, storing, and analyzing topographic and soil features, vegetative and land-use patterns, and human interventions for management planning on GIS; and
- Linking spatially distributed data sets to computer models to simulate the effects of alternative land-use strategies on the flows of agricultural and natural resources.

With the availability of more comprehensive data bases and improvements to computational facilities, watershed managers in India are actively involved in planning, implementing, and operating more effective and responsive watershed management programs. A major goal of watershed management programs in the Indian Subcontinent is to reduce environmental degradation to permissible limits, while increasing biomass production to optimum levels (Chandra and Bhatia, 2000). Work accomplished to attain this goal is helping watershed managers and decision-makers achieve sustainable production and use of all natural resources on watershed lands. However, a large variability in the levels of environmental degradation, types of biomass production possible, and interactions of the controlling hydrologic and biological processes on these lands exist. Therefore, watershed managers must consider these issues in their planning efforts.

Since independence, integrated watershed management and water resource development has been extensive in India. However, many experiences are the outcome of sectoral, short-term, and goal-oriented programs that are largely incompatible with sustainability or that have been insufficiently pursued and lack acceptance by the local people (Chandra and Bhatia, 2000). To achieve sustainability of India's natural resources, a shift in watershed management strategies to more people-oriented programs is necessary.

2.2. VIEWS IN THE UNITED STATES

The United States has a long history of watershed management, which was largely a local concern in the first half of the 19th century (Adams et al., 2000; Neary, 2000). Citizens of towns and villages petitioned their local governments for permission to construct hydropower plants for mills and irrigation channels and to develop privately owned water-supply systems. Eastern and Midwestern cities with political power and financial resources condemned large tracts of land to develop water-supply reservoirs. These cities usually acted unilaterally because no other level of government or segment of society claimed authority over water resources. Water rights became the provenance of territorial and later, state governments in the Western and other arid parts of the country. A general scarcity of water in this region lead to the enactment of laws to define water rights and settle disputes over watershed resources, with the prevailing ideas that "first in time was first in right."

At the end of the 19th century, a series of disastrous floods in the Eastern United States prompted citizens to request that the federal government enact flood-control programs (Adams et al., 2000). In response, the U.S. Army Corps of Engineers was directed to build structures that harnessed rivers to protect life and property from flooding. Legislation was passed that extended federal authority over all navigable waters and prohibited construction of any structure that modified the flow of any waterway without recommendation from the Corp's Chief Engineer and authorization from the Secretary of War.

The early 20th century saw increased activity in watershed management at the municipal level. Public-works departments were building water-supply reservoirs, constructing drinking-water systems, and installing sewage-treatment plants, while private power companies were constructing hydropower dams (Adams et al., 2000). Beginning in the 1900s and continuing through the 1960s, legislation addressing a diversity of water-related issues was enacted at the federal level (see Chapter 3). Construction of hydropower dams, development of irrigation supplies, and flood-control measures were the focus of this legislation. Some legislation authorized the initiation of needed water-supply and delivery programs, while other legislation was promulgated to stimulate the country's economy during economic slumps.

Concerns about cost-efficiency and interagency battles led to efforts to coordinate the watershed management policies and programs of the federal agencies (Fairchild, 1993; Adams et al., 2000). The Water Resources Planning Act of 1965 established a national Water Resources Council and many regional river-basin commissions to develop more effective watershed-related policies and provide financial assistance state-level water planning. Interstate commissions were also established to coordinate water-supply and flood- and sewage-control districts, and the activities of state water resource agencies and the federal government.

By the early 1980s, most of the efforts to coordinate the activities of the federal agencies had failed because many decision-makers refused to acknowledge the authority of the interstate commissions and other regional entities (Adams et al., 2000). States also saw the federal efforts as attempts to bypass their role in watershed planning. The Water Resources Council was abolished in 1981. Simultaneously, the federal government largely abandoned its river-basin planning efforts. However, in the early 1990s, the inadequacy of the federal government to engage local entities in nonpoint source pollution control and ecological watershed restoration prompted (Adams et al., 2000) the U.S. Environmental Protection Agency (EPA) to call for a watershed management approach to environmental planning (U.S. Environmental Protection Agency, 1991).

Based on its history of watershed management , the United States has learned that:

- Public opinion and local, state, and federal laws and regulations concerning watershed management differ from east to west and from small towns to large cities. These differences exist largely because of the perceived scarcity of water and other natural resources and the availability of financial resources (Adams et al., 2000; Eckman et al., 2000).
- Integrated and comprehensive watershed and river-basin management occurs when local, state, and federal management and regulatory agencies work together. Agencies at all levels of government must have the financial resources and policymaking authority to effectively develop watershed resources.
- When federal agencies are the primary decision-makers, local and state entities, both public and private, often resist watershed-related determinations. Decentralized, participatory approaches to planning efforts, currently being emphasized in the United States, are sensitive to the interests of a wide range of stakeholders and recognize that compromises are necessary when attempting to maximize production and minimize environmental impact. Additionally, participatory planning introduces effective conflict-management approaches to watershed management from a sustainability perspective.

We present watershed management perspectives, problems, and programs in the Lake States and Northeast, the Southeast, the Pacific Northwest, and the Colorado River Basin for a perspective on the lessons learned. Regional case studies illustrate the wide range of viewpoints of watershed management in the United States.

2.2.1. Lake States, Northeast, and East

The Lake States (Minnesota, Wisconsin, and Michigan) and the states of the Northeast (Connecticut, New York, Massachusetts, Vermont, New Hampshire, Maine) are collectively the most heavily populated regions of the United States. Many of the largest megalopolis centers— the Twin Cities of Minneapolis and St. Paul, Detroit, Boston, New York, Milwaukee, and Chicago— are in this region. Diverse landscapes of intermingling forests and woodlands, agricultural lands, and expanding urban developments require watershed management perspectives, problems, and programs diversity. Land-use changes in the region alter river-flow dynamics and result in channel bank erosion. In 1902, J. T. Rothrock, the first Director of Pennsylvania's Department of Forestry (Verry et al., 2000), noted these activities. Twenty years later, Raphael Zon, Director, USDA Forest Service Lake States Forest Experiment Station, reinforced the concept of land and water interactions in his treatise on "Forests and Water in the Light of Scientific Investigation" (Zon, 1927). Since completion of this dissertation, a watershed management approach has been the foundation of the land stewardship efforts in the Lake States and Northeast.

2.2.1.1. The Setting

Intermingling forested, agricultural, and urban landscapes characterize the Lake States and Northeast. Interspersed across the region are thousands of lakes, ponds, rivers, and streams. Mostly mixed-conifer forests—many publicly managed—are scattered in the northern latitudes. Hardwood forests in private and public ownerships occupy the southern areas of the region. European settlers cleared most of the original forests to grow a variety of agricultural food crops and to raise livestock. Many current forest structures are second-growth stands that succeeded the earlier heavily harvested and often burned original stands. The region's developing cities, towns, and villages are encroaching onto these agricultural lands as farmers sell their property to urban developers and leave the agricultural sector. People in the Lake States and Northeast enjoy comparatively high personal incomes and increasing amounts of leisure time—conditions that accelerate the demands on the region's natural resources.

2.2.1.2. Major Watershed Management Problems

Watershed research efforts in the Lake States and Northeast reflect the region's watershed management problems. Studies of nutrient cycling and streamflow chemistry have been part of watershed research since the middle 1960s (Bormann and Likens, 1979; Edwards and Helvey, 1991; Hornbeck and Swank, 1992; Likens and Bormann, 1995). Results from this research were used to evaluate the long-term impacts of timber harvesting, atmospheric deposition, and other human-caused disturbances on soil-water leaching losses and on the responses of streamflow chemistry to these disturbances (Verry et al., 2000). Historical records of precipitation and streamflow are useful to study the trends of watershed responses to planned and unexpected disturbances (Driscoll et al., 1989; Edwards and Helvey, 1991; Adams et al., 1993). For example, regional controls of sulfur emissions have resulted in decreasing concentrations of sulfate in precipitation and streamflow in the White Mountains of New Hampshire. Nitrogen and calcium in streamflows originating in the Appalachian

Mountains of West Virginia have been gradually increasing due to nitrogen saturation from human inputs of nitrogen. Studies at the Marcell Experimental Forest in north-central Minnesota on the interactions of upland and wetland nutrient cycling (Verry and Timmons, 1982) illustrate the significance of atmospheric chemistry on streamflow and watershed biota chemistry.

Results of water-yield improvement experiments at the Marcell Experiment Forest, the Fernow Experimental Forest in central West Virginia, the Hubbard Brook Experimental Forest in New Hampshire, and the Leading Ridge Experimental Forest near Pennsylvania State University (Hornbeck et al., 1993) will help sustain the municipal water supplies in the region. Immediate increases in annual water yields were observed on these study sites after the removal of forest overstories. However, these increases rapidly diminished without subsequent regrowth control. The prolonged increases in posttreatment water yields exhibited in the Southeast and Colorado River Basin do not occur in the Lake States and Northeast (Verry et al., 2000). Shallow soils and rooting depths, a short growing season, low evapotranspiration rates, rapid root occupancy and leaf-area development, and a complete recharge of soil moisture in the dormant season limits the magnitude and duration of water-yield increases in the Lake States and Northeast.

Maintaining watershed lands in a productive condition is a critical issue in the region. Protecting watersheds and channel conditions by reducing erosion and sediment movement from roads, trails, and stream crossings has been the objective of many watershed research efforts. Most studies have been conducted on the Fernow Experimental Forest, where the construction and evaluation of minimum-standard roads have been the research focus (Kochenderfer and Helvey, 1987; Helvey and Kochenderfer, 1988; Kochenderfer, 1995). When stream channels change from their "natural" state, they become unstable and are unable to transport water and debris from their watershed without causing excessive erosion and sedimentation (Rosgen, 1996; Verry, 2000). Research has provided some information about how channel-forming streamflows are related to and caused by land-use practices, however, more information is needed to assess fully the region's "desirable" stream conditions.

2.2.1.3. Watershed Management Programs

We summarize 3 watershed management programs in the Lake States and Northeast to show how results from nearly 50 years of watershed research have been incorporated into a watershed management approach to land stewardship. In one program, watershed managers on the White Mountain National Forest have incorporated research results from the Hubbard Brook Experimental Forest (Likens et al., 1970; Martin et al., 1984); Federer et al., 1989; Hornbeck et al., 1993, 1997) and elsewhere in the Northeast into their planning process and landscape management programs (Verry et al., 2000). Nutrient-depletion tables for calcium, magnesium, and other base cations and a variety of forest disturbances and timber harvesting methods have been used to select silvicultural practices on this forest to provide the optimum protection of site-nutrient capitals and forest productivity. Researchers and managers are applying research findings from the Hubbard Brook Experimental Forest to determine whether, when, and where to impose restrictions on timber harvesting practices that accelerate base cation losses and elevate nitrates in the streamflow. On the White Mountain National Forest, research programs on snow hydrology and water quality have helped managers understand the impacts of ski-area construction and snowmaking on hydrology, erosion and sedimentation, and water quality.

Another program at the Minnesota Department of Natural Resources is using a composite of watershed research findings to respond to the provisions of the Clean Water Act, its amendments, and other related laws. These efforts have led to issuance of best management practices (BMPs) in Minnesota. These practices are continuously reviewed, tested, and improved to provide guidance and make recommendations for road construction, placement of filters between roads and streams, applications of prescribed fire, and use of pesticides to protect the quality of the state's water resources (Verry et al., 2000). Recently, efforts have considered riparian areas, wildlife habitats, forest-soil productivity, and cultural resources. A restriction to focus on site-specific guidelines at the landscape level is currently being considered.

The Chesapeake Bay, which runs north from the coast of Virginia, is the largest and, because of its shallowness, most productive estuary in the United States. Land-use practices on the surrounding watershed largely determine water quality, aquatic habitat vitality, and ultimately, the health and resilience of the bay itself (Swanson, 1994; Verry et al., 2000). However, by the 1970s, the cumulative effects of more than 200 years of intense urban and industrial development, loss of forests, increasing pollution and runoff, and accumulations of sediments, nutrients, and industrial wastes resulted in declines in fisheries, shellfish, waterfowl, and bay grasses. Concerned citizens, scientists, and managers agreed that restoration of the Chesapeake Bay was needed . In 1983, the states of Pennsylvania, Maryland, and Virginia, and the District of Columbia and the Chesapeake Bay Commission and local federal agencies collaborated in an EPA study, which is the third watershed management program we summarize. These entities formed a watershed partnership (see Chapter 3) to develop and implement a comprehensive and highly coordinated program to improve and protect the land, water, and other natural resources of the Chesapeake Bay. This partnership has grown into a unique multi-state and multiagency effort that coordinates the bay-related activities of hundreds of federal, state, local, and intergovernmental agencies working with dozens of private business and civic and environmental organizations for long-term regional stewardship.

2.2.2. Southeast

The Southeast United States has abundant natural resources and highly diverse, attractive ecosystems. This region experiences demographic shifts, job opportunities, and other sociocultural factors that contribute to a rapidly growing range of public interests and often, changing views of how land, water, and other natural resources should be managed. This situation is most complex in the Appalachian Mountain region where mixed land-ownerships, intense population pressures, and natural resource diversity challenge watershed managers (Swank and Tilley, 2000). Past and present watershed research efforts and management experiences in the Southeast provide information about the decision-making process used to meet the population's natural resource requirements. The Coweeta Hydrologic Laboratory, in the Appalachian Mountains of western North Carolina, has a history of interdisciplinary research and management that is directly applicable to the decision-making process. This cooperative Forest Service program is the case study that we examine.

2.2.2.1. *The Setting*

The coastal plains, the Piedmont area, and mountain ranges characterize the Southeast. Forests cover 870,000 km^2 of which 810,000 km^2, which is classified as timberland[4], account for about 40% of the productive timberland in the United States. Most of the timberland in the region is held in nonindustrial private ownership. Other timberlands are on National Forest System land, administered by other public agencies, or owned by forest industries. Upland and other hardwood types grow on about 50% of the timberland, and pine plantations grow on almost one-third of the remaining timberland.

The Southeast has emerged as a leader in the forest-products industry in the past decade—accounting for 35% of the solid wood products and 25% of the paper production worldwide, and 50% of the softwood and about 40% of the hardwood timber harvest in the United States (Wear and Bolstad, 1998; Wear et al., 1998; Swank and Tilley, 2000). Cotton and tobacco remain economically important components of the agricultural sector.

The region contains an abundance of water and wildlife resources and a diversity of recreational opportunities. Originating in the southern Appalachians mountains, parts of 73 major watersheds and 9 major rivers provide drinking water to the Southeast's major cities and metropolitan areas. The metropolitan hub of the region is Atlanta, Georgia. Other centers of commerce are New Orleans, Louisiana; Miami and St. Petersburg, Florida; Charlotte, North Carolina; and Charleston, South Carolina.

2.2.2.2. *Major Watershed Management Problems*

Water quantity is not a serious problem in the Southeast, although the cumulative impacts of land use on downstream water-quality variables are a concern (Bolstad and Swank, 1997). The major problem confronted by watershed managers in the region is integrating water, timber production, hunting and fishing, and hiking into comprehensive, ecosystem-based, multiple-use land and natural resource management programs (Swank and Tilley, 2000). Sometimes the conflicting uses of these natural resources and the required land-use patterns must be reconciled with continuously increasing population pressures and maintenance of the region's unique environmental qualities. A related problem is providing access to the natural resource areas. Roads for timber harvest, recreational facilities, and remote sites of interest must be planned, constructed, and continually maintained considering the impact on other natural resources. An ecological approach to achieve broader multiple-use objectives in watershed management is necessary to resolve these issues.

2.2.2.3. *Watershed Management Programs*

One of the earliest demonstrations of multiple-use management in the Southeast was implemented at the Coweeta Hydrologic Laboratory in 1962. At the time, substantial

---

[4] Forestland where tree species traditionally used for industrial wood products make up at least 10% of the stocking and produce or are capable of producing crops of industrial wood and are not withdrawn from timber utilization.

controversy existed over the Multiple-Use Sustained Yield Act, passed by the United States Congress in 1960, because "on-the-ground" examples of the multiple-use concept were lacking. Therefore, prescriptions incorporating water, timber production, and recreation into multiple-use watershed management were prepared, and the prescribed treatments were implemented in the Coweeta basin (Hewlett and Douglass, 1968). Douglass and Swank (1972, 1976) and Swank et al. (1998) presented analyses of the responses of the natural resources to the treatments imposed therefore, we do not describe them in this chapter. Generally, the analyses showed that water yields increased; stormflow discharges, sediment loads, and nutrient losses were minimal or had negative responses; and vegetation, wildlife, and hunting and hiking values increased.

The 35-year old watershed-based studies at Coweeta demonstrate that the southern Appalachian forests can be successfully managed for a variety of uses (Swank et al., 1994; Swank, 1998; Swank and Tilley, 2000). While some conflicts among uses were reported, the ecosystem changes observed were reversible and, consequently, opportunities remain to alter future management strategies to meet future goals. Many findings from Coweeta have been factored into ongoing forest planning and management in the Southeast. These findings also show that a watershed is appropriate for making, applying, and evaluating management-oriented decisions. A watershed provides an on-the-ground framework where managers, conservation and environmental groups, and policymakers can collaborate on issues concerning land, water, and other natural resource management.

In 1992, the need for an ecological approach to achieve a broader perspective of watershed management led to establishment of an ecosystem-based, multiple-use pilot program in the Wine Spring Creek Basin of western North Carolina—about 50 km west from Coweeta. Identifying, implementing, and evaluating management practices that achieve desired watershed conditions, consistent with the ecosystem-management philosophy of the Forest Service (Kessler et al., 1992; Thomas, 1996) and sound land stewardship, is the core of this program (Swank and Tilley, 2000). Existing forest and natural resources management plans are used to define the desired future natural-resource conditions and to specify the prescriptions to achieve these conditions (Figure 2.2). Prescribed watershed management practices include stand-replacement prescribed burning to restore degraded hardwood-pine communities, stimulate forage production, and promote oak regeneration; silvicultural treatments to regenerate oak and increase structural diversity; and stream-habitat improvements on impoverished aquatic habitats. Researchers, managers, and citizens have formed a partnership to collaborate on the program's decision-making process and to promote participation by all concerned stakeholders.

2.2.3. Pacific Northwest

Watershed management in the Pacific Northwest over the past century has been inconsistent. Knowledge of past watershed management practices, their occurrences through time, their spatial distribution, and their impacts on land, water, and other natural resources are largely incomplete (Beschta, 2000). We do know that millions of cubic meters of old-growth timber have been removed from the region's forest to meet economic and social demands, and that hundreds of thousands of herbivores previously foraged on the rangeland-watersheds of the region. Unfortunately, information about the ecological and environmental consequences of the harvesting and grazing is lacking. The effects of increased

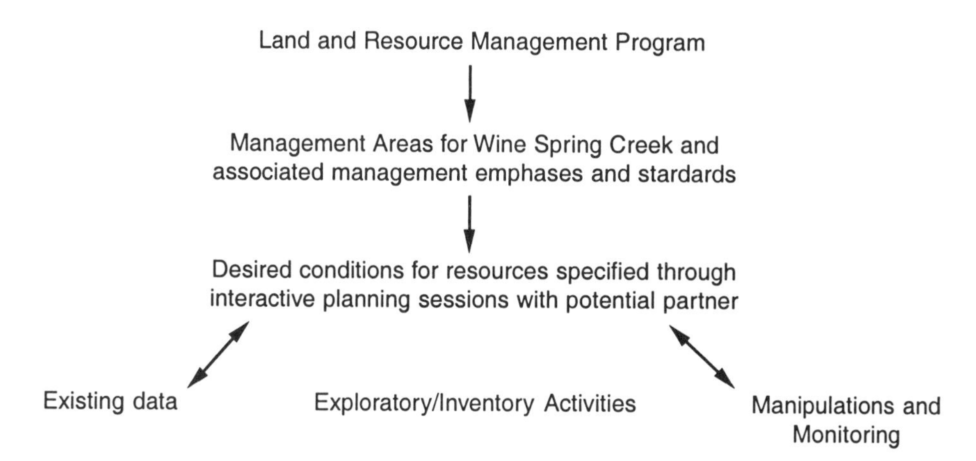

Figure 2.2. Planning for ecosystem-based, multiple-use management on the Wine Creek Watershed in western North Carolina (adapted from Swank and Tilley, 2000). Through participatory planning efforts, existing forest and natural resource management plans define the desired conditions of resources and subsequent research and management perspectives.

urbanization and industrial activities on land, water, and other natural resources have also been inadequately evaluated. Therefore, increased understanding of the pressures inherent in society's use of watershed-based resources is necessary for those living in the Pacific Northwest to progress in their land stewardship efforts.

2.2.3.1. The Setting

Landscapes of the region include the moist temperate forests and agricultural lands westward from the tops of the coastal mountain ranges to the Pacific Ocean; the drier coniferous forests eastward from the tops of the coastal mountain ranges and throughout the scattered mountains inland; and the rolling Palouse prairies of eastern Washington, Oregon, and Idaho. The Columbia River Basin, one of the largest of the Pacific Northwest watersheds, encompasses the area that is upriver of the Columbia estuary, including a significant portion that extends northward into Canada (Hessburg et al., 1999; Beschta, 2000). Within the biophysical regions of the Columbia are many smaller watersheds of varying areas.

Besides its historical position as a leading United States timber producer, the region also generates and exports large quantities of hydropower from the series of dams and reservoirs along the Columbia River. This dam and reservoir system also regulates the river for navigation, provides irrigation water, and serves flood-control purposes. Seattle, Washington, and Portland, Oregon, are the population centers of commerce in the region.

2.2.3.2. Major Watershed Management Problems

Over the past century, exploration, development, and the constant use of the natural resources in the Pacific Northwest have caused many watershed problems that are still being confronted. Historically, agriculturalists used streams until they dried-up, since maintenance of instream habitats to protect water quality and aquatic habitats was not an acknowledged beneficial use (Beschta, 2000; Mann and Plummer, 2000). In the past, ranchers competing for decreasing forage resources on watershed lands compensated by increasingly stocking larger livestock numbers, often leading to a "tragedy of the commons."[5] Recently, lowering the livestock stocking on many public rangelands has mitigated this problem.

In the Pacific Northwest, improved harvesting technologies facilitates access to timber on steep mountain watersheds, which often leads to excessive soil losses and sedimentation increases (Sidle, 1998; 2000). Additionally, conversion of forestlands to agriculture or urban development has contributed to the wetland loss (National Research Council, 1995). Land conversions and other improper management practices on many watershed lands has resulted in nonpoint and point sources of potential pollution. These and other long-term, continuing human activities increase concerns about the detrimental and cumulative effects to the environment in the region.

2.2.3.3. Watershed Management Programs

The forests, rangelands, and rivers of the Pacific Northwest have been transformed in ways unimaginable only decades ago. The region's watershed lands, which are part of these "inherited landscapes," are managed to help satisfy a modern society's economic, social, and environmental demands (Beschta, 2000; Galliano and Loeffler, 2000). The federal government manages much of the land in the Pacific Northwest and, as such, the administrative policies and management practices are different from those on the intermingled privately owned lands. Nevertheless, stakeholder collaboration is necessary to develop effective watershed management programs because watersheds follow topographic rather than political boundaries.

Management programs on watersheds draining into the ocean carefully consider the region's Pacific salmon species. These anadromous fish migrate thousands of kilometers of streams, rivers, estuaries, and oceans to complete their life cycle. The Pacific salmon is important economically and culturally, and their status indicates the environmental conditions of riparian and aquatic habitats. While in freshwater environments, the Pacific salmon synthesize information on a complex of management, political, social, and biological factors that influence the sustainability of their habitats (Meehan, 1991; Stouder et al., 1997). Other indicators of the status of streamside watershed resources include water temperatures, sediment levels, and flow alterations.

Recently, changing timber harvest practices in the Pacific Northwest has influenced the planning and implementation of the region's watershed management programs. Providing multistoried, late-successional forested habitats, such as those preferred by the northern spotted owl, have curtailed timber harvesting on the region's extensive tracts of forested watershed lands (Marcot and Thomas, 1997; Christensen et al., 1999; Hicks et al., 1999). The reduction in the volume of trees with commercial value that are cut and removed from a forest has caused drastic shifts in watershed management priorities. Prompted by concern about the region's riparian ecosystem health, regulations have been established by the federal

[5] When individuals use limited natural resources without limit to pursue their own best interest.

government and the states in the region regarding tree retention, timber harvest, shade retainment, and the management of snags along stream channels (Hairston-Strang and Adams, 2000). Watershed managers no longer work in relative isolation on narrowly defined issues. Instead, their environment provides for interdisciplinary efforts to synthesize research results and incorporate management experiences from a wide range of technical and social topics.

2.2.4. Colorado River Basin

The perennial shortage of water has conditioned and circumscribed the lives of the inhabitants of the Colorado River Basin. Early settlers quickly appropriated the limited, highly variable surface water supplies. Later, people used electrical pumps to reach groundwater reserves for increasing agriculture and urbanization needs (Baker and Ffolliott, 2000). Today, because most of the water supplied to this rapidly growing region is pumped from underground basins, regional water tables are steadily declining, which adversely affects local economies. Increasingly high water costs have forced landowners to convert many square kilometers of agricultural land into housing developments. In this region, retention of productive agricultural land now requires "imported" water from sources such as the Central Arizona Project, which transports water from the Colorado River to central and southern Arizona for agricultural and urban use.

Water previously used for agriculture and the remaining surface and groundwater resources are incapable of sustaining the growth of regional municipalities and industries in the Colorado River Basin. In fact, anticipated increasing demands for water are expected to cause widespread shortages of relatively inexpensive water—local shortages already occur. Holistically planned watershed management programs can mitigate the magnitude of these shortages, while meeting the demands for other natural resources.

2.2.4.1. The Setting

The Colorado River Basin—most of Arizona and portions of New Mexico, Colorado, Wyoming, Utah, Nevada, and California—drains through nearly 650,000 km^2 of land before entering the Gulf of California. The area is divided into the Upper and Lower Basins at Lee's Ferry, Arizona, which is about 15 km south of the Utah-Arizona border. The Upper Basin contains an area of 285,000 km^2, while the Lower Basin is 365,000 km^2 in size.

A cyclic climatic pattern of winter rain and snow at higher elevations, a prolonged spring drought, a summer rainfall season, and a fall drought characterize this basin (Baker, 1999). Precipitation averages 400 mm annually in the Upper Basin, with the heaviest concentrations in the mountains (Hibbert, 1979; Baker and Ffolliott, 2000). The Lower Basin receives an average of 330 mm of annual precipitation (Hibbert, 1979; Baker and Ffolliott, 2000). Precipitation that becomes streamflow is more than 5 times greater in the Upper Basin (16%) than in the Lower Basin (3%). However, precipitation and streamflow in both basins vary considerably from year-to-year. Seasonally, streamflow is concentrated in a few months each year during snowmelt.

Vegetation on the mostly publicly administered watersheds of the Upper Basin includes forests of spruce-fir, lodegepole pine, Douglas-fir, and ponderosa pine intermixed with quaking aspen at the higher elevations; pinyon-juniper woodlands, mountain brush lands, and big sagebrush at intermediate elevations; and other shrublands and grasslands at the lower elevations. Riparian ecosystems are in narrow bands of trees, shrubs, and herbaceous plants along streams. Irrigated agriculture is in the valleys and plains.

Vegetation in the Lower Basin includes mixed conifer and ponderosa pine forests at the higher elevations; pinyon-juniper woodlands and chaparral shrublands at intermediate elevations; and desert shrublands and grasslands at the lower elevations. Riparian ecosystems consist of narrow corridors of vegetation along streams that drain into the Colorado River and its tributaries. Irrigated agriculture occurs in the valleys. The large number of people settling in the Colorado River Basin—specifically in the large metropolitan areas of Denver, Colorado; Albuquerque, New Mexico; Phoenix and Tucson, Arizona; and their suburbs—has made this one of the fastest growing regions in the United States. The basin enjoys relatively high incomes, low unemployment, and increasing amounts of leisure time (Ffolliott et al., 2000)—conditions that typically accelerate the demand for natural resources.

2.2.4.2. *Major Watershed Management Problems*

Without additional water importation, saline water conversion, cloud-seeding advancement, and rigorously enforced conservation measures, people living in the Colorado River Basin will continue to rely on variable surface water and diminishing groundwater reserves (Baker and Ffolliott, 2000). To increase water yields from the basin's forests, woodlands, and shrublands—while sustaining other natural resource use at optimal levels—the federal government has been studying the long-term effects of vegetative manipulations on streamflows that originate on upland watersheds (Gary, 1975; Leaf, 1975; Sturges, 1975; Martinelli, 1975; Baker, 1999). Unfortunately, environmental concerns and recent court rulings have restricted most large-scale vegetative manipulations by prohibiting or restricting tree cutting and timber harvest in the region. This situation has constrained the watershed management options on many public lands in the basin. Other watershed management concerns in the Colorado River Basin include further deterioration of fragile soil and water resources, sustaining the flows of high-quality water within established limitations, and the declining health of some watershed lands due to improper hydrologic functioning (Ffolliott et al., 2000).

2.2.4.3. *Watershed Management Programs*

Early watershed management programs emphasized that water was a limited commodity in the Colorado River Basin. It is not surprising, therefore, that watershed management programs from the early 1940s through the beginning of the 1980s focused on improving the region's water supplies by increasing water yields on upland watersheds through vegetation management. Clearcutting, other silvicultural treatments, and conversions from high to low water-consuming vegetation were implemented on many watersheds to increase water yields and sustain timber, forage, and wildlife resources (Hibbert, 1979; Baker, 1999; Baker and Ffolliott, 2000). Successful additional water yields were attributed mostly to decreases in evapotranspiration.

Passage of the National Environmental Policy Act, the Clean Water Act, and creation of the EPA and similar state-level agencies helped to increase the public's awareness of environmental policy issues regarding natural resource management. By the late 1970s, those responsible for watershed management programs began to examine the physical, chemical, and biological qualities of water originating on upland watersheds and the potential sources of pollution affecting these water qualities—these concerns remain the current focus (Ffolliott et al., 1997).

Watershed management programs in the Colorado River Basin at the beginning of the 21st century incorporate the interrelationships among land use, soil, and water, and the relationship between upland and downstream areas (Lopes and Ffolliott, 1992; Baker et al., 1995; Ffolliott et al., 2000). Practices include reducing the adverse impacts to soil and water resources to maintain watersheds that are in good condition and sustaining the flows of high-quality water originating on upland watersheds. Water shortages, always a concern in the region, will likely become more acute as the human population continues to increase. To increase productivity, rehabilitating watersheds and riparian systems that are in poor condition is another a priority of watershed management programs in the Colorado River Basin (Baker, 1999). Rehabilitation efforts focus on restoring the proper hydrologic functioning on the degraded lands. Careful implementation of BMPs will help to achieve these goals (see Chapter 5).

Current watershed management programs in the Colorado River Basin represent a holistic approach to managing the biological, physical, and sociocultural elements on landscapes that watershed boundaries delineate. Reducing the demands for watershed resources by using these limited resources more efficiently will foster integrated management of these elements. Additionally, watershed management in the basin is considering the wildland urban interface[6] as anticipated increasing development in this area will escalate demands on increasingly limited and impacted resources.

2.3. SUMMARY

To achieve sustainable flows of natural resources and agricultural benefits from watershed lands, a diversity of watershed management programs continue to be initiated by national governments, bilateral and multilateral agencies, and nongovernment organizations. These benefits are not solely dependent on the physical and biological characteristics of the watersheds. The cultural backgrounds and economy of the effected urban and rural populations and the nature of the involved governments and other institutions must also be integrated into viable watershed management programs to meet program objectives. Through case studies in Mexico, Taiwan, Jordan, and India, we examined how these factors are globally interrelated. Watershed management perspectives, problems, and programs in the United States were viewed in case studies from the Lake States and Northeast, the Southeast, the Pacific Northwest, and the Colorado River Basin. Collectively, these case studies show the many ways that incorporating watershed management into the conservation, sustainable development, and use of land, water, and other natural resources is possible. Examination of the case studies may also suggest future direction to achieve better land stewardship through watershed management.

---

[6] The line, area, or zone where structures and other human development meet or intermingle with undeveloped wildland or vegetative fuels.

REFERENCES

Adams, C., Noonan, T., Newton, B. "Watershed Management in the 21st Century: National Perspective." In *Land Stewardship in the 21st Century: The Contributions of Watershed Management*, P. F. Ffolliott, M. B. Baker, Jr., C. B. Edminster, M. C. Dillon, and K. L. Mora, tech. coords. Fort Collins, CO: Rocky Mountain Research Station, USDA Forest Service, Proceedings RMRS-P-13, 2000, pp. 21-29.

Adams, M. B., Edwards, P. J., Wood, F., Kochenderfer, J. N. Artificial watershed acidification on the Fernow Experimental Forest, USA. Journal of Hydrology 1993; 150:505-519.

Agarwal, A. "Development in Hydrological Drought Studies." In *Hydrological Developments in India Since Independence*. Roorkee, India: National Institute of Hydrology, 1992, pp. 289-311.

Aguilar, A. G., Ezcurra, E., Garcia, T., Mazari-Hiriart, M., Pisaty, I. "The Basin of Mexico." In *Regions at Risk: Comparisons of Threatened Environments*, J. X. Kasperson, R. E. Kasperson, and B. L. Turner II, eds. Tokyo: United Nations University, 1995, pp. 305-549.

Baker, M. B., Jr., compiler, *History of Watershed Research in the Central Arizona Highlands*. Fort Collins, CO: USDA Forest Service, Rocky Mountain Research Station, General Technical Report RMRS-GTR-29, 1999.

Baker, M. B., Jr., Ffolliott, P. F. "Contributions of Watershed Management Research to Ecosystem-Based Management in the Colorado River Basin." In *Land Stewardship in the 21st Century: The Contributions of Watershed Management*, P. F. Ffolliott, M. B. Baker, Jr., C. B. Edminster, M. C. Dillon, and K. L. Mora, tech. coords. Fort Collins, CO: Rocky Mountain Research Station, USDA Forest Service, Proceedings RMRS-P-13, 2000, pp.117-128.

Baker, M. B., Jr., DeBano, L. F., Ffolliott, P. F. "Hydrology and Watershed Management in the Madrean Archipelago." In *Biodiversity and Management of the Madrean Archipelago: The Sky Islands of Southwestern United States and Northwestern Mexico*, L. F. DeBano, P. F. Ffolliott, A. Ortega-Rubio, G. J. Gottfried, R. H. Hamre, and C. B. Edminster, tech. coords. Fort Collins, CO: Rocky Mountain Forest and Range Experiment Station, USDA Forest Service, General Technical Report RM-GTR-264, 1995, pp. 329-337.

Beschta, R. L. "Watershed Management in the Pacific Northwest: The Historical Legacy." In *Land Stewardship in the 21st Century: The Contributions of Watershed Management*, P. F. Ffolliott, M. B. Baker, Jr., C. B. Edminster, M. C. Dillon, and K. L. Mora, tech. coords. Fort Collins, CO: Rocky Mountain Research Station, USDA Forest Service, Proceedings RMRS-P-13, 2000, pp. 109-116.

Bhatia, K. K. S. "Man's Influence on Hydrologic Cycle." In *Hydrological Developments in India Since Independence*. Roorkee, India: National Institute of Hydrology, 1992, pp. 413-440.

Bojorquez Tapia, L. A., Ezcurra, E., Mazari-Hiriart, M., Diaz, S., Gomez, P., Alcantar, G., Megarejo, D. "Basin of Mexico: A History of Watershed Mismanagement." In *Land Stewardship in the 21st Century: The Contributions of Watershed Management*, P. F. Ffolliott, M. B. Baker, Jr., C. B. Edminster, M. C. Dillon, and K. L. Mora, tech. coords. Fort Collins, CO: Rocky Mountain Research Station, USDA Forest Service, Proceedings RMRS-P-13, 2000, pp. 129-137.

Bolstad, P. V., Swank, W. T. Cumulative impacts of landuse on water quality in a southern Appalachian watershed. Journal of the American Water Resources Association 1997; 33: 519-533.

Bormann, F. H., Likens, G. E., *Pattern and Process in a Forested Ecosystem*. New York: Springer-Verlag, 1979.

Brooks, K. N., Eckman, K. "Global Perspective of Watershed Management." In *Land Stewardship in the 21st Century: The Contributions of Watershed Management*, P. F. Ffolliott, M. B. Baker, Jr., C. B. Edminster, M. C. Dillon, and K. L. Mora, tech. coords. Fort Collins, CO: Rocky Mountain Research Station, USDA Forest Service, Proceedings RMRS-P-13, 2000, pp. 11-20.

Brooks, K. N., Ffolliott, P. F., Gregersen, H. M., DeBano, L. F., *Hydrology and the Management of Watersheds*. Ames, IA: Iowa State University Press, 1997.

Brooks, K. N., Ffolliott, P. F., Gregersen, H. M., Easter, K. W., *Policies for Sustainable Development: The Role of Watershed Management*. Washington, DC: U.S. Department of State, EPAT Policy Brief 6, 1994.

Brooks, K. N., Gregersen, H. M., Ffolliott, P. F., Tejwani, K. G. "Watershed Management: A Key to Sustainability." In *Managing the World's Forests: Looking for Balance Between Conservation and Development*, N. P. Sharma, ed. Dubuque, IA: Kendall/Hunt Publishing Company, 1992, pp. 455-487.

Chandra, S., *Long Range Forecasting of Hydrological Aspects of Drought for Tropical and Subtropical Regions*. Geneva, Switzerland: Secretariat of the World Meteorological Organization, Technical Reports in Hydrology and Water Resources 35, 1993.

Chandra, S., Bhatia, K. K. S. "Water and Watershed Management in India: Policy Issues and Priority Areas for Future Research." In *Land Stewardship in the 21st Century: The Contributions of Watershed Management*, P. F. Ffolliott, M. B. Baker, Jr., C. B. Edminster, M. C. Dillon, and K. L. Mora, tech. coords. Fort Collins, CO: Rocky Mountain Research Station, USDA Forest Service, Proceedings RMRS-P-13, 2000, pp. 158-165.

Cheng, J. D., Hau, H. K., Ho, J., Chen, T. C. "Watershed Management for Disaster Mitigation and Sustainable Development in Taiwan." In *Land Stewardship in the 21st Century: The Contributions of Watershed Management*, P. F. Ffolliott, M. B. Baker, Jr., C. B. Edminster, M. C. Dillon, and K. L. Mora, tech. coords. Fort Collins, CO: Rocky Mountain Research Station, USDA Forest Service, Proceedings RMRS-P-13, 2000, pp. 138-148.

Christensen, H. H., Raettig, T. L., Sommer, P., tech. eds., *Northwest Forest Plan: Outcomes and Lessons Learned From the Northwest Ecosystem Adjustment Initiative*. Portland, OR: Pacific Northwest Research Station, USDA Forest Service, General Technical Report PNW-GTR-484, 1999.

Douglass, J. E., Swank, W. T., *Streamflow Modification Through Management of Eastern Forests*. Asheville, NC: Southeastern Forest Experiment Station, USDA Forest Service, Research Paper SE-94, 1972.

Douglass, J. E., Swank, W. T. Multiple use in southern Appalachian hardwoods - a ten year case history. Proceedings of the 16th International Union of Forestry Research Organization's World Congress. Vienna, Austria: IUFRO Secretariat, 1976, pp. 425-436.

Driscoll, C. T., Likens, G. E., Hedin, L. O., Eaton, J. S., Bormann, F. H. Changes in the chemistry of surface waters. Environmental Science and Technology 1989; 23: 137-143.

Dutton, R. W., Clarke, J. I., Battikhi, A. M., *Arid Land Resources and Their Management: Jordan's Desert Margin*. London: Kegan Paul International, 1998.

Eckman, K. E., Gregersen, H. M., Lundgren, A. L. "Watershed Management and Sustainable Development: Lessons Learned and Future Directions." In *Land Stewardship in the 21st Century: The Contributions of Watershed Management*, P. F. Ffolliott, M. B. Baker, Jr., C. B. Edminster, M. C. Dillon, and K. L. Mora, tech. coords. Fort Collins, CO: Rocky Mountain Research Station, USDA Forest Service, Proceedings RMRS-P-13, 2000, pp. 37-43.

Edwards, P. E., Helvey, J. D. Long-term ionic increases from a central Appalachian forested watershed. Journal of Environmental Quality 1991; 20:250-255.

Ezcurra, E., Mazari-Hiriat, M. Are megacities viable? A cautionary tale from Mexico City. Environment 1996; 38(1):6-15, 26-35.

Fairchild, W. D. "A Historical Perspective on Watershed Management in the United States." In *Proceedings Watershed 93: A National Conference on Watershed Management*. Cincinnati, OH: USEPA Center for Environmental Publications, 1993, pp. 5-10.

Federer, C. A., Hornbeck, J. W., Trillon, L. M., Pierce, R. S. Long-term depletion of calcium and other nutrients in eastern US forests. Environmental Management 1989; 13:593-601.

Ffolliott, P. F., Baker, M. B., Jr., Lopes, V. L. "Watershed Management in the Southwest: Past, Present, and Future." In *Land Stewardship in the 21st Century: The Contributions of Watershed Management*, P. F. Ffolliott, M. B. Baker, Jr., C. B. Edminster, M. C. Dillon, and K. L. Mora, tech. coords. Fort Collins, CO: Rocky Mountain Research Station, USDA Forest Service, Proceedings RMRS-P-13, 2000, pp. 30-36.

Ffolliott, P. F., DeBano, L. F., Strazdas, L. A., Baker, M. B., Jr., Gottfried, G. J. Hydrology and water resources: A changing emphasis? Hydrology and Water Resources in Arizona and the Southwest 1997; 27:65-66.

Galliano, S. J., Loeffler, G. M., *Scenery Assessment: Scenic Beauty at the Ecoregion Scale*. Portland, OR: Pacific Northwest Research Station, USDA Forest Service, General Technical Report PNW-GTR-472, 2000.

Gary, H. L., *Watershed Management Problems and Opportunities for the Colorado Front Range Ponderosa Pine Zone: The Status of Our Knowledge*. Fort Collins, CO: Rocky Mountain Forest and Range Experiment Station, USDA Forest Service, Research Paper RM-139, 1975.

Gowda, K. N., Jayaramaiah, K. M. "Watershed Approach - A Ray of Hope to Rainfed Farmers." In *Watershed 2000: Science and Engineering Technology for the New Millennium*, M. Flug and D. Frevert, eds. Reston, VA: Environment & Water Resources Institute, American Society of Civil Engineers, 2000. [CD-ROM] Windows

Hairston-Strang, A. B., Adams, P. W. Riparian management area condition for timber harvests conducted before and after the 1994 Oregon water protection rules. Western Journal of Applied Forestry 2000; 15:147-153.

Haque, M. C. "Flood Hydrology." In *Hydrological Developments in India Since Independence*. Roorkee, India: National Institute of Hydrology, 1992, pp. 151-163.

Helvey, J. D., Kochenderfer, J. N. Culvert sizes needed for small drainage areas in central Appalachians. Northern Journal of Applied Forestry 1988; 5:123-127.

Hessburg, P. F., Smith, B. G., Kreiter, S. D., and others, *Historical and Current Forest and Range Landscapes in the Interior Columbia River Basin and Portions of the Klamath and Great Basins. Part 1: Linking Vegetation Patterns and Landscape Vulnerability to Potential Insect and Pathogen Disturbances*. Portland, OR: Pacific Northwest Research Station, USDA Forest Service, General Technical Report PNW-GTR-458, 1999.

Hewlett, J. D., Douglass, J. E., *Blending Forest Uses*. Asheville, NC: Southeastern Forest Experiment Station, USDA Forest Service, Research Paper SE-37, 1968.

Hibbert, A. R., *Managing Vegetation to Increase Flow in the Colorado River Basin.* Fort Collins, CO: Rocky Mountain Forest and Range Experiment Station, USDA Forest Service, General Technical Report RM-66, 1979.

Hicks, L. L., Stabins, H. C., Herter, D. R. Designing spotted owl habitat in a management forest. Journal of Forestry, 1999; 97(7):20-25.

Hornbeck, J. W., Swank, W. T. Watershed ecosystem analysis as a basis for multiple-use management of eastern forests. Ecological Applications 1992; 2:238-247.

Hornbeck, J. W., Martin, C. W., Eagar, C. Summary of water yield experiments at Hubbard Brooks Experimental Forest, New Hampshire. Canadian Journal of Forest Research 1997; 27:2043-2052.

Hornbeck, J. W., Adams, M. B., Corbett, E. S., Verry, E. S. Lynch, J. A. Long-term impacts of forest treatments on water yield: A summary for northeastern USA. Journal of Hydrology 1993; 150:323-344.

Jha, M. K., Haque, M. E., Sinha, C. P. A critique of drought definitions and indicators. Water Management Journal 1993; 2:27-37.

Kessler, W. V., Salwasser, H., Cartwright, C. W., Jr., Caplan, J. A. New perspectives for sustainable natural resources management. Ecological Applications 1992; 2:221-225.

Kochenderfer, J. N., *Using Open-Top Pipe Culverts to Control Surface Water on Steep Road Grades.* Radnor, PA: Northeastern Forest Experiment Station, USDA Forest Service, General Technical Report NE-194, 1995.

Kochenderfer, J. N., Helvey, J. D. Using gravel to reduce soil losses from minimum-standard forest roads. Journal of Soil and Water Conservation 1987; 42:46-50.

Kumar, C. P. "Ground Water Modeling." In *Hydrological Developments in India Since Independence.* Roorkee, India: National Institute of Hydrology, 1992, pp. 235-261.

Leaf, C. F., *Watershed Management in the Rocky Mountain Subalpine Zone: The Status of Our Knowledge.* Fort Collins, CO: Rocky Mountain Forest and Range Experiment Station, USDA Forest Service, Research Paper RM-137, 1975.

Likens, G. E., Bormann, F. H., *Biogeochemistry of a Forested Ecosystem.* New York: Springer-Verlag, 1995

Likens, G. E., Bormann, F. H., Johnson, N. M., Fisher, D. W., Pierce, R. S. Effects of forest cutting and herbicidal treatment on nutrient budgets in the Hubbard Brook watershed-ecosystem. Ecological Monographs 1970; 40:23-47.

Lopes, V. L., Ffolliott, P. F. "Hydrology and Watershed Management of Oak Woodlands in Southeastern Arizona." In *Ecology and Management of Oak and Associated Woodlands: Perspectives in the Southwestern United States and Northern Mexico,* P. F. Ffolliott, G. J. Gottfried, D. A. Bennett, V. M. Hernandez C., A. Ortega-Rubio, and R. H. Hamre, tech. coords. Fort Collins, CO: Rocky Mountain Forest and Range Experiment Station, USDA Forest Service, General Technical Report RM-218, 1992, pp. 71-77.

Mann, C. C., Plummer, M. L. Can science rescue salmon. Science 2000; 289:716-719.

Marcot, B. G., Thomas, J. W., *Of Spotted Owls, Old Growth, and New Policies: A History Since the Interagency Scientific Committee Report.* Portland, OR: Pacific Northwest Research Station, USDA Forest Service, General Technical Report PNW-GTR-408, 1997.

Martin, C. W., Noel, D. S., Federer, C. A. Effects of forest clearcutting in New England on stream chemistry. Journal of Environmental Quality 1984; 13:204-210.

Martinelli, M., Jr., *Water-Yield Improvement From Alpine Areas: The Status of Our Knoledge.* Fort Collins, CO: Rocky Mountain Forest and Range Experiment Station, USDA Forest Service, Research Paper RM-138, 1975.

Meehan, W. P., *Influences of Forest and Rangeland Management on Salmonid Fishes and Their Habitats.* Bethesda, MD: American Fisheries Society, 1991.

Mishra, G. C., Singh, S. K. "Ground Water and Conjunctive Use." In *Hydrological Developments in India Since Independence.* Roorkee, India: National Institute of Hydrology, 1992, pp. 165-194.

National Research Council, *Wetlands: Characteristics and Boundaries.* Washington, DC: Committee on the Characterization of Wetlands, National Resources Council, National Academy of Science, 1995.

Neary, D. G. "Changing Perceptions of Watershed Management From a Retrospective Viewpoint." In *Land Stewardship in the 21st Century: The Contributions of Watershed Management,* P. F. Ffolliott, M. B. Baker, Jr., C. B. Edminster, M. C. Dillon, and K. L. Mora, tech. coords. Fort Collins, CO: Rocky Mountain Research Station, USDA Forest Service, Proceedings RMRS-P-13, 2000, pp. 167-176.

Rosgen, D., *Applied River Morphology.* Pagosa Springs, CO: Wildland Hydrology, 1996.

Shahbaz, M., Sunna, B. "Integrated Stufies of the Azraq Basin in Jordan." In *Land Stewardship in the 21st Century: The Contributions of Watershed Management,* P. F. Ffolliott, M. B. Baker, Jr., C. B. Edminster, M. C. Dillon, and K. L. Mora, tech. coords. Fort Collins, CO: Rocky Mountain Research Station, USDA Forest Service, Proceedings RMRS-P-13, 2000, pp. 149-157.

Sidle, R. C. Progress in evaluating the impacts of timber harvesting on soil mass movement. International Seminar on Watershed Conservation and Sustainable Management; 1998 April 22-23; Tiachung, Taiwan: National Chung Hsing University, 1998, pp. 1-8.

Sidle, R. C. "Watershed Challenges for the 21st Century: A Global Perspective for Mountainous Terrain." In *Land Stewardship in the 21st Century: The Contributions of Watershed Management*, P. F. Ffolliott, M. B. Baker, Jr., C. B. Edminster, M. C. Dillon, and K. L. Mora, tech. coords. Fort Collins, CO: Rocky Mountain Research Station, USDA Forest Service, Proceedings RMRS-P-13, 2000, pp. 45-56.

Stouder, D. J., Bisson, P. A., Naiman, R. J., eds., *Pacific Salmon and Their Ecosystems: Status and Future Options*. New York: Chapman & Hall, 1997.

Sturges, D. L., *Hydrologic Relations on Undisturbed and Converted Big Sagebrush Lands: The Status of Our Knowledge*. Fort Collins, CO: Rocky Mountain Forest and Range Experiment Station, USDA Forest Service, Research Paper RM-140, 1975.

Swank, W. T. "Multiple Use Forest Management in a Catchment Context." In *Multiple Land Use and Catchment Management*, M. Cresser and K. Pugh, eds. Aberdeen, Scotland, UK: The Maccaulay Land Use Institute, 1998, pp. 27-37.

Swank, W. T., Tilley, D. R. "Watershed Management Contributions to Land Stewardship: Case Studies in the Southeast." In *Land Stewardship in the 21st Century: The Contributions of Watershed Management*, P. F. Ffolliott, M. B. Baker, Jr., C. B. Edminster, M. C. Dillon, and K. L. Mora, tech. coords. Fort Collins, CO: Rocky Mountain Research Station, USDA Forest Service, Proceedings RMRS-P-13, 2000, pp. 93-108.

Swank, W. T., McNulty, S. G., Swift, L. W., Jr. "Opportunities for Forest Hydrology Applications to Ecosystem Management." In *Forest Hydrology: Proceedings of the International Symposium*, T. Ohta, Y. Fukushima, and M. Suzuki, eds. Tokyo, Japan: International Union of Forestry Organizations, Organizing Committee of the International Symposium on Forest Hydrology, 1994, pp. 19-29.

Swanson, A. P. Watershed restoration in the Chesapeake Bay. Journal of Forestry 1994; 92(8):37-38.

Thomas, J. W. Forest Service perspective on ecosystem management. Ecological Applications 1996; 6:703-705.

Thorud, D. B., Brown, G. W., Boyle, B. J., Ryan, C. M. "Watershed Management in the United States in the 21st century." In *Land Stewardship in the 21st Century: The Contributions of Watershed Management*, P. F. Ffolliott, M. B. Baker, Jr., C. B. Edminster, M. C. Dillon, and K. L. Mora, tech. coords. Fort Collins, CO: Rocky Mountain Research Station, USDA Forest Service, Proceedings RMRS-P-13, 2000, pp. 57-64.

Tung, Ching-Pin, T., Hong, Nien-Ming. "Sustainability Indicator for Water and Land Resources. In *Watershed 2000: Science and Engineering Technology for the New Millennium*, M. Flug and D. Frevert, eds. Reston, VA: Environment & Water Resources Institute, American Society of Civil Engineers, 2000. [CD-ROM] Windows

U.S. Environmental Protection Agency, *The Watershed Protection Approach: An Overview*. Washington, DC: U.S. Environmental Protection Agency, EPA 503/9-92/002, 1991.

Verry, E. S. "Water Flow in Soils Sustaining Hydrologic Function." In *Riparian Management in Forests of the Continental Eastern United States*, E. S. Verry, J. W. Hornbeck, and C. A. Dolloff, eds. Boca Raton, FL: Lewis Publishers, 2000, pp. 99-124.

Verry, E. S., Timmons, D. R. Waterborne nutrient flow through and upland-peatland watershed in Minnesota. Ecology 1982; 63:1456-1467.

Verry, E. S., Hornbeck, J. W., Todd, A. H. "Watershed Research and Management in the Lake States and Northeastern United States." In *Land Stewardship in the 21st Century: The Contributions of Watershed Management*, P. F. Ffolliott, M. B. Baker, Jr., C. B. Edminster, M. C. Dillon, and K. L. Mora, tech. coords. Fort Collins, CO: Rocky Mountain Research Station, USDA Forest Service, Proceedings RMRS-P-13, 2000, pp. 81-92.

Wear, D. N., Bolstad, P. Land use changes in southern Appalachian landscapes: Spatial analysis and forest evaluation. Ecosystems 1998; 1:575-594.

Wear, D. N., Abt, R., Mangold, R. People, space and time: Factors that will govern forest sustainability. Transactions of the North American Wildlife and Natural Resources Conference 1998; 63:348-361.

Zon, R. 1927. Forests and water in the light of scientific investigation. In *U.S. Natural Waterways Commission: Final Report.* Senate Document 469, Congress, 2d Session, V:205-302.

3

A RETROSPECTIVE VIEW

A retrospective view of watershed management in the United States and globally is helpful to understand the issues confronted in land stewardship (see Chapter 4) and the contributions of watershed management to future land stewardship (see Chapter 5). Participants at the conference on "Land Stewardship in the 21st Century: The Contributions of Watershed Management" reviewed the following 4 general topics, which are also discussed in the chapter, from a retrospective viewpoint (Ffolliott et al., 2000).

- A history of watershed management—A perspective of how watershed management has broadened in recent years;
- Lessons learned—Past research findings and management experiences in watershed management demonstrate the effects of vegetative manipulations and other land-use practices on streamflow regimes, erosion and sedimentation processes, water quality constituents, and other natural resources;
- Emerging tools and technologies—Advances in existing computer technology and the emergence of new tools and technologies that capture and analyze spatial data sets have helped managers to evaluate the cumulative effects of watershed management practices and improve the understanding of watershed processes; and
- Locally-led initiatives in watershed management—The significance of incorporating sociocultural perspectives into planning and implementing watershed management programs for better land stewardship is exemplified by the development of watershed management partnerships, councils, and corporations between the public and private sectors and the creation of citizen-based advisory organizations.

3.1. HISTORY OF WATERSHED MANAGEMENT

Watershed management has evolved from a focus on water resource management and the hydrologic cycle to the current integrated approach of managing the biological, physical, and social elements on a landscape defined by watershed boundaries. A brief history of watershed management—mostly from a United States' perspective—is presented below.

3.1.1. Early History

Management of watershed lands is rooted in the history of human civilization. Verdic texts from 1,000 B.C. indicate that the people of present-day India understood the importance of the hydrologic cycle, from which the modern science of hydrology and watershed management evolved (Chandra, 1990). Earlier, the development of cities around the Mediterranean Sea and throughout the Middle East depended on effective water management, including the construction of sustainable wells and aqueducts, to bring water from its source to the urban people (Neary, 2000). However, sustainable water management and the application of hydrologic engineering for this purpose declined sharply with the collapse of the Roman Empire and entry of western civilization into the Dark Ages. Hundreds of years passed before watershed management skills emerged. During and following the Renaissance, observation, measurement, and experimentation of water resources expanded throughout Europe.

Native cultures in the western hemisphere, including the Huari and Tihuanaco empires of present-day Peru and Bolivia and the Inca and Mayan civilizations, made significant achievements in watershed management. Irrigation canals, agricultural terracing, subterranean sewage and drainage systems, and indoor running water and toilet facilities supported the burgeoning populations. The Hohokam people, occupying the Sonoran Desert of southern Arizona from 600 to 1,200 A.D., developed extensive networks of irrigation canals to almost present-day engineering standards (Reid and Whittlesey, 1997). A contemporary history of watershed management follows.

3.1.2. Late 19th Century

Water deprivation influenced the Mormon views of water and management of watershed lands in Utah's Great Salt Lake Valley in the late 19th century. Arriving to find a desert landscape next to a salt sea, the Mormons launched into extensive water-development projects that resulted in nearly 25,000 km^2 of irrigated agriculture land in Utah and several surrounding states by the end of century (Neary, 2000). Through much of the 20th century, the Mormon experiences and technologies influenced the water development programs accomplished by the U.S. Bureau of Reclamation in the Western United States. Water and watershed management gained a further foothold in the United States with the creation of forest reserves (later national forests) by the U.S. Congress. The Pettigrew Amendment to the 1897 Sundry Civil Appropriations Bill specified that forest reserves could be established only to, ". . . improve and protect the forest within the reservation, or for the purpose of securing favorable conditions of water flows . . ." (Steen, 1976). By the turn of the century, the focus of watershed management within the context of forestry was centered on water supply and flood prevention.

3.1.3. Early 20th Century

The early 20th century was the beginning of intensive research programs to refine and enhance earlier watershed management practices. In 1903, near Emmental, Switzerland, the first experimental watersheds were established to determine the influence of forest cover on the water economy of the region (Hibbert, 1967). Unfortunately, ensuring that the observed streamflow differences between the 2 watersheds studied were caused solely by differences in forest cover was impossible.

The control-watershed approach[7] was used for the first time in the Wagon Wheel Gap study of 1911 (Bates and Henry, 1928; Hibbert, 1967). After 8 years of calibration, one of the Wagon Wheel Gap watersheds, located high in the Colorado Rockies, was denuded of forest cover and, for 7 more years, the streamflow from the denuded watershed was compared with that from the control watershed, with a forest cover (Bates and Henry, 1928; Hibbert, 1967). While the regression techniques of later years were unavailable to evaluate the treatment effect, the researchers demonstrated that removing a forest cover caused streamflow to increase.

Many legislative actions affected watershed management practices in the United States in the early 20th century (Neary, 2000). The Reclamation Act of 1902 increased settlement of large tracts of public land in the Western United States through construction of federally financed reservoir and irrigation canals and other watershed management projects. The Weeks Law of 1911 recognized the value of vegetation on watershed lands by authorizing the President to, "... reserve any part of the public lands wholly or in part covered with timber and undergrowth, whether commercial or not, as public reservations." The Weeks Law further acknowledged that poorly managed watersheds could increase the likelihood of flood flows and produce fluvial and riparian damage. Extensive tracts of forests standing at the time of European settlement in the Eastern United States had been reduced to barren, logged-over shrublands that were in poor hydrologic condition by the middle of the 1920s. In responding to this situation, the Clark McNary Act of 1924 offered incentives to state and private landowners to encourage restoration of their logged-over forests to improve timber production and watershed protection.

3.1.4. Middle 20th Century

The middle 20th century saw increased activity and large-scale investments in watershed management by the federal agencies of the United States, principally the Bureau of Reclamation, the Army Corps of Engineers, the Soil Conservation Service, and the USDA Forest Service. The Forest Service was especially active in watershed management through its diverse national forests manage programs established to meet society's needs for natural resources and to rehabilitate abandoned and eroded lands (Neary, 2000). The Forest Service also made large investments in watershed management research at the Coweeta Hydrologic Laboratory in North Carolina, San Dimas Experimental Forest in California, Sierra Ancha Experimental Forest in Arizona, Fernow Experimental Forest in West Virginia, Fraiser Experimental Forest in Colorado, H. J. Andrews Experimental Forest in Oregon, Hubbard Brook Experimental Forests in New Hampshire, and the Beaver Creek watersheds in Arizona.

A key legislative action by the U.S. Congress was the Flood Control Act of 1936, which mandated the Soil Conservation Service (now the Natural Resources Conservation Service) to conduct watershed management programs in upstream areas to reduce the incidence of flooding. This act asserted federal responsibility for flood control on the nation's navigable rivers and their tributaries, which lead to the Corps of Engineers building large structural

[7] Comparing the streamflow from 2 similar watersheds throughout a calibration period by treating one watershed and leaving the other untreated as a control.

engineering projects to control floods and erosion on downstream portions of large river basins. The Watershed Protection and Flood Prevention Act of 1954 created a small watershed restoration and management program that provided incentives to private and public landowners to maintain or improve soil productivity and reduce destructive flooding. The Multiple-Use Sustained Yield Act of 1960, which remains integral to national forest management, was the first piece of legislation to mention the major uses of watershed lands—wood, water, forage, range, wildlife, and recreation—in one federal law (Steen, 1976). The Multiple-Use Sustained Yield Act explicitly states that national forest management does not have a natural-resource priority but that land, water, and other natural resources should be sustainably managed.

Concurrent with enactment of this cornerstone legislation was a changing perception about the discipline of watershed management. In his milestone textbook *Forest Influences*, Kittredge (1948) used the 1944 Society of American Foresters definition of watershed management—"the administration and regulation of the aggregate resources of a drainage basin for the production of water and the control of erosion." Kittredge's concept of watershed management for water and erosion control prevailed. Later, Coleman (1953) stated that watershed management should focus on manipulating the vegetative cover on watershed lands to alter the hydrologic processes and achieve control over water yields. Pavari (1962) felt that the objective of watershed management in Europe was to protect soil, climate, and water resources to meet the need for wood, water, and other products and to secure the full use of all lands in the country's general interest. Dixon (1964) and Ogrosky and Mockus (1964) took an engineering view in Chow's *Handbook of Applied Hydrology* (1964a) when addressing the physical aspects of watershed management without considering biological or ecological factors.

Ed Dortignac, head of the Water Resources Branch of the USDA Forest Service in the 1960s, felt that watershed management was on the threshold of a great opportunity in land stewardship because inadequate water supplies—previously only a problem in the dryland regions of the United States—were now an issue in the humid Eastern United States (Dortignac, 1967). Dortignac stated that prescriptions using wood, forage, wildlife, and recreation resources that improved water yields and controlled, maintained, or improved soil stability could provide effective watershed management. Dortignac's view of watershed management in 1967 was similar to the multiple-use philosophy of the middle 1990s.

3.1.5. Late 20th Century

By the late 20th century, the availability of land, water, and other natural resources was becoming increasingly constrained. One-half of the countries worldwide had low-to-very-low fresh water supplies (Postel, 1992; The World Bank, 1993; Brooks et al., 1997). Wood resources, livestock forage, wildlife, and recreational opportunities necessary for sustaining or improving people's lives were dwindling. Ecological and economic disasters on regional, national, and global scales were mounting. This situation increased recognition of the significance of and need for a holistic, ecosystem-based, multiple-use approach to land stewardship (Eckman et al., 2000; Neary, 2000). Watershed management history at the end of the 20th century had evolved into this perspective.

3.2. LESSONS LEARNED

Knowledge about the relationship between the vegetative cover on watershed lands and the flows of water from these watersheds has been significantly enhanced in the last 50 years. Many research studies and extensive management experiences have shown that vegetative manipulations on watershed lands can alter streamflow regimes, erosion and sedimentation processes, water quality, and other natural resources. This knowledge continues to be incorporated globally into land, water, and other natural resource management. The information obtained and lessons learned from these studies provides a foundation when considering the contributions of watershed management to future land stewardship.

3.2.1. Streamflow

Watershed management studies in the United States and elsewhere in the world have provided long-term, high-quality measurement of precipitation and resultant streamflow regimes. Estimates of how much water the vegetative cover returns to the atmosphere through evapotranspiration have also been obtained (Brooks and Ffolliott, 1996; Ffolliott and Brooks, 1996; Megahan and Hornbeck, 2000). Studies of throughfall[8] have allowed the separation of the canopy interception within the evapotranspiration component of a water budget. Investigations of ground-level precipitation infiltration and percolation have helped to differentiate water-flow pathways and the amount of streamflow potentially available. Process studies have identified source-areas on which streamflow originates. With this knowledge, prescribed treatments on experimental watersheds show how disturbances change key hydrologic processes and relationships, such as tree harvesting, conversion of plant species, and fire.

3.2.1.1. Annual-water Yields

In summarizing the results from experimental watershed studies throughout the world, Hibbert (1967), Bosch and Hewlett (1982), Troendle and King (1985), Hornbeck et al. (1993), Whitehead and Robinson (1993), and others arrived at the following 3 generalizations about annual-water yields.

- Reduction of forest cover increases streamflow volumes;
- Establishment of forest cover on sparsely vegetated watersheds decreases streamflow volumes; and
- Responses to treatments are variable, relatively short-term, and unpredictable (Table 3.1).

Magnitudes of increases in annual-water yields (first generalization) have been highly variable, leading to the third generalization. Bosch and Hewlett (1982) felt that they could mitigate the unpredictability of streamflow to treatment responses through application of reliable computer simulators to predict changes in annual-water yields following vegetative treatments. Several computer simulators have modeled these changes (Brooks et al., 1997; Guertin et al., 2000).

Most experimental watershed studies indicate that the responses of streamflow to vegetative treatments also depend on the amount of precipitation falling on the watershed

[8] Precipitation that directly reaches the ground.

(Hewlett, 1967; Hornbeck et al., 1993; Baker, 1999; Frasier and Holland, 2000). Therefore, it is reasoned that precipitation below a threshold amount is effectively used by residual vegetation on the watershed and increases in herbaceous plant covers (Box 3.1). Under similar precipitation regimes at or above these thresholds, increases in streamflow are roughly proportional to the reductions in forest, woodland, and shrubland densities (Megahan and Hornbeck, 2000). Several studies have also shown that streamflow increases following partial tree cutting, conversion treatments, and fire is often related to the configurations or locations of the treatment (Troendle, 1983; Hornbeck et al., 1993; Baker, 1999).

3.2.1.2. Flow Distribution

While annual water-yield changes can occur with vegetative treatments on watersheds, in the short-term, it is important to know how streamflow changes are distributed over time

Box 3.1 Minimum Precipitation is Necessary

An analysis by Hibbert (1979) showed that vegetative manipulations in the Colorado River Basin could increase annual water yields only on watersheds receiving more than 480 mm of annual precipitation. He reasoned that the vegetation left on the watershed, including increases in the understory biomass, consumes precipitation below this amount following the treatments. This suggests that high-elevation mixed conifer and ponderosa pine forests and some low-elevation chaparral shrublands have the best theoretical potentials for increasing annual water yields through regional vegetation management.

Table 3.1. Potential annual water-yield increases by vegetative management on forested uplands and rangelands in the Western United States.

Vegetative Type	Annual Water Yield Increase (mm)	References
Forest		
Subalpine	25-150	Leaf, 1975; Hibbert, 1979
Mixed conifer	75-100	Rich and Thompson, 1974; Hibbert, 1979
Aspen	100-150	Hibbert, 1979
Ponderosa Pine	25-165	Hibbert, 1979; Baker, 1999
Pinyon-juniper	0-10	Hibbert, 1979; Baker, 1999
Rangeland		
Mountain brush	25-75	Johnson et al., 1969; Hibbert, 1979
Chaparral	100	Hibbert, 1979; Baker, 1999
Sagebrush	0-12	Sturges, 1975; Hibbert, 1979
Semi-desert shrublands	negligible	Hibbert, 1979

(Megahan and Hornbeck, 2000). A differentiation between peak flows[9] and flood flows[10] should be made when considering the effects of vegetative treatments on flow distribution. Bankfull is the stage where water completely fills the channel system without spreading onto the adjacent flood plain (Dunne and Leopold, 1978; Brooks et al., 1997).

Magnitudes of peak flows can be increased, decreased, or remain unchanged after cutting trees on a watershed (Harr, 1986; Wright et al., 1990; Megahan and Hornbeck, 2000). Causes of increases in peak flows from vegetative treatments include wet, hydrologically responsive soils resulting from decreased evapotranspiration losses after tree cutting, soil compaction, and road construction. Whether or not a change occurs depends on what part of the hydrologic system is altered, to what degree, and the permanence of the alteration. Fire can also affect flow distributions. Low-intensity prescribed burning has little effect on flood flows, but intense wildland fire can significantly increase flood flows (Lull and Reinhart, 1972; Campbell et al., 1977; DeBano et al., 1998). Increased soil-water repellence from intense burning is often a significant factor leading to increased flood flows. However, even if flood flows increase on small watersheds because of intense wildland fire, detecting these effects on large river basins is unlikely because of the increased channel storage along larger streams (Chow, 1964b).

If peak flows do not increase following tree cutting or wildland fire, the occurrence of low flows[11] will rise if annual-water yields expand. Most streamflow increases occur at low flow, or they are associated with augmented baseflow[12] or delayed flow (Hornbeck et al., 1997; Megahan and Hornbeck, 2000). While flood flows do not usually increase due to vegetative manipulations, increases in the duration of streamflow near the bankfull stage can lead to bank and streambed erosion in channels that are susceptible to these problems (Van Haveren, 1988; Troendle et al., 1988).

3.2.2. Erosion

Erosion is a normal and continuous geologic process that varies through time in response to changing climates and site conditions. Erosion rates on undisturbed watersheds in good condition are usually small. However, both surface and mass erosion can increase to unacceptable levels when natural (extreme storm events, wildland fire) or human (tree cutting, road construction) disturbances occur (Dunne and Leopold, 1978; Brooks et al., 1997; Megaham and Hornbeck, 2000). Soil erosion can reduce upland productivity and adversely affect aquatic environments and other beneficial uses of surface water.

3.2.2.1. Surface Erosion

Surface erosion[13] has always been a critical concern on agricultural lands because of its impact on site productivity. However, surface erosion has recently become a concern on

[9] The maximum flows resulting from a runoff event.
[10] Flows that exceed the stream channel capacity as defined by the bankfull stage.
[11] Flows that are lower than the expected average flows.
[12] Water that drains from a watershed to sustain streamflow during dry periods. Baseflow sources are groundwater flows, which move out of an underground reservoir of saturated material, the upper surface of which is a water table. Another source of baseflow is the slow and continuous drainage from unsaturated soil, which can be sufficient to sustain streamflow for long periods in the absence of a water table.
[13] The movement of soil particles by raindrop impact or overland water flow.

wildland watersheds as timber-harvesting rotations and other land-use practices intensify. Studies show that increased surface erosion following tree cutting is confined to severely disturbed and compacted sites on harvested landscapes and, as a consequence, is limited mostly to skid trails[14], landings[15], and roads (Megaham and Kidd, 1972; Martin, 1988). Erosion rates have been predicted from simulation models—the Revised Universal Soil Loss Equation or the Water Erosion Prediction Project—with varying levels of success (Tiscareno-Lopes et al., 1995; Yoder and Lown, 1995; Brooks et al., 1997; Elliot, 2000). A variety of best management practices address increased surface erosion from timber harvesting activities (Brown, 1980; Burroughs and King, 1989; Brown et al., 1993; Martin and Hornbeck, 1994) and other land-use practices (see Chapter 5).

3.2.2.2. Mass Erosion

Shallow landslides (debris flows, avalanches, and torrents) or deep-seated landslides (slumps and earthflows) are common forms of mass erosion[16] that occur on watershed lands. Studies have shown that tree cutting can increase the risk of shallow landslides by reducing root strength and increasing soil-water content (Sidle et al., 1985; Brooks et al., 1997; Sidle, 2000). Simulation models based on Geographic Information Systems (GIS) (see below) have been developed to evaluate the risk of these landslides occurring on landslide-prone terrain following timber harvesting (Montgomery and Dietrich, 1994).

Roads in mountainous areas are also problematic for mass erosion (Megahan and Kidd, 1972; Megahan and Hornbeck, 2000) because of:

- Exposure of erodible soil and subsoil resulting from their construction;
- Reduced infiltration on the road surface;
- Increased gradients on cut and fill slopes; and
- Concentrated overland flow from precipitation excess and interception of subsurface flow.

Preventing timber harvest on steep, unstable slopes and eliminating the permanent conversion of forest to sparse herbaceous cover can mitigate mass erosion (Dunne and Leopold, 1978; Brooks et al., 1997). Avoiding road networks in high-hazard areas and carefully designing, constructing, and maintaining roads can reduce mass erosion on roads.

3.2.3. Sedimentation

Sedimentation[17] is a complementary natural process to erosion. However, on-site erosion does not necessarily equate to downstream sediment yield because of storage en route. The differences between erosion rates and sediment yield, often quantified by the sediment delivery ratio, determine the effects of long-term sediment storage on a watershed (Brooks et al., 1997; Megahan and Hornbeck, 2000). Land-use conditions affect the sediment delivery

---

[14] The pathways of movement along the ground of trees or tree segments from where they were cut to other areas.
[15] Where trees or tree segments are loaded onto trucks and transported from the forest to a mill for processing.
[16] The movement of aggregates of soil particles en mass in response to gravitional force.
[17] The transfer of eroded soil materials downslope to stream channels and then downstream through the drainage system.

ratio, texture of the eroded soil material, and local stream environment (Dunne and Leopold, 1978; Brooks et al., 1997). As the size of the watershed increases, the sediment delivery ratio generally decreases.

Increased sedimentation can cause a variety of environmental problems (Megahan and Hornbeck, 2000). For example, the U.S. Environmental Protection Agency (1992) found that sedimentation impairs a greater length of streams in the United States than any other pollutant including nutrients, pathogens, pesticides, and dissolved oxygen. Sedimentation concerns also affect fishery values. High concentration of suspended sediments can damage fish gills and the well-being of aquatic insects. Increased bedload sediments can interfere with fish spawning success. Fine organic and clay-sized lithic sediments can also act as vectors for the downstream transport of pesticides, organic chemicals, or heavy metal.

3.2.3.1. Sediment Delivery from Surface Erosion

Sediment from diffuse sources of surface erosion—for example, sites where tree cutting has occurred, a road fill exists, or sites following wildland fire—moves only short distances downslope if there are no concentrated sources of runoff (Rich, 1962; Megahan and Ketcheson, 1996; Brooks et al., 1997). However, sediment from road culverts and other concentrated runoff sources can travel much further. The distance of sediment travel from these sources (Packer, 1967; Swift, 1986; Megahan and Ketcheson, 1996) depends on the:

- Occurrence of obstructions on and gradients of the slope below the road;
- Volume of erosion; and
- Amount of overland water flow.

Sediment volume deposited generally decreases exponentially downslope; therefore, most is stored near the source. Because surface erosion selects smaller soil particles, the sediments produced from surface erosion consist of clays, silts, and fine sands.

3.2.3.2. Sediment Delivery from Mass Erosion

Sediment delivery from mass erosion is usually higher in quantity than that from surface erosion. Debris-avalanche landslides usually occur at the heads of steep drainages on sites of high-water concentration, and they follow the drainage-path down to the lower channel system (Sidle, 1985; Megahan and Hornbeck, 2000). Unlike the smaller soil particles supplied to stream channels from surface erosion, the sediment delivered to channels by landslides can range from clay to boulders. Landslide activity is not always detrimental, however. For example, sediment from landslides following wildland fires in the coastal mountain range of Oregon is essential to the sustainability of aquatic habitats in the stream channels (Benda and Dunne, 1997).

3.2.3.3. Sediment Transport in Streams

Some eroded soils are eventually transported to streams. Most of the smaller sediments (clays and silt) move rapidly through the channel system as wash load (Dunne and Leopold, 1978; Brooks et al., 1997; Megaham and Hornbeck, 2000). Larger sediments move as bedload materials and can have varying residence times depending on their particle size. Sediment-transport events are usually more episodic in dryland areas than in humid areas because of the frequently encountered intermittent streamflow regimes in dryland areas (Heede et al., 1988). Urbanization often increases the transport of sediment because of construction activities.

A potentially important loss of nutrients (nitrogen, potassium, calcium, etc.) and heavy metals (iron, zinc, copper, etc.), which is often ignored, is transported from a watershed or river basin by sediment (Fisher and Mickley, 1978; Gosz et al., 1980). This loss can result in depletion of nutrients on a watershed. Therefore, sedimentation cumulative effects[18] are a possibility when eroded soil materials move through a stream system.

3.2.4. Water Quality

Sediment has long been a water-quality parameter of interest to watershed managers. Since the 1960s, however, nutrient cycling, the impact of introduced chemicals, and water temperature have also been included when considering water quality (Satterland and Adams, 1992; Brooks et al., 1997; Megahan and Hornbeck, 2000). Because pollutant transportation in streams depends on water movement, any land-use activity that alters the volumes or timing of runoff will affect the rates of nutrient, chemical, and other pollutant transport. Herbicide applications on upland sites also affect water quality.

3.2.4.1. Nutrients

Nutrient-cycling studies that began in the middle 1960s at the Hubbard Brook Experiment Forest have become important additions to watershed experiments at many other locations in the United States and worldwide. This holistic approach to the evaluation of watershed experiments led to the introduction of what has been called watershed ecosystem analysis (Hornbeck and Swank, 1992). The premise of a watershed ecosystem analysis is that many physical, chemical, and biological processes operating within a watershed ecosystem are interrelated. Results from the nutrient-cycling studies at Hubbard Brook and elsewhere allow values to be attached to the physical, chemical, and biological constituents of concern. This provides a basis for assessing the impacts of natural or human-related changes on the parameters (Hornbeck and Swank, 1992; Megahan and Hornbeck, 2000). A watershed ecosystem analysis is used to evaluate how individual or combinations of land-uses might affect the nutrient cycles of a watershed and, in turn, the health and productivity of the ecosystem.

The climate, mineral weathering, soil characteristics, vegetation, biological processes, and natural and human disturbances affect nutrient leaching on watersheds supporting natural vegetation. Studies have shown that watersheds covered by natural vegetation and free of recent disturbances generally possess nutrient cycles with relatively little loss of nutrients from the site and baseline concentrations of nutrients in stream water are frequently low (Hornbeck et al., 1997). However, sediment particles also transport nutrients from a watershed (Gifford and Busby, 1973; Angino et al., 1974; Gosz et al., 1980). Nevertheless, nutrient concentrations are seldom a water-quality issue on these types of watersheds, although the nutrient capital present on the watershed's stream biota can be a consideration.

---

[18] Environmental changes influenced by a combination of land-use activities.

Watershed ecosystem analysis has contributed to the study of atmospheric deposition on watershed lands. Research shows that acidic depositions in the industrialized Eastern United States can lower the pH of streams and lakes and mobilize inorganic aluminum to levels that are toxic to aquatic biota (Cronan and Schofield, 1990). High concentrations of mineral acids in precipitation can also deplete the base cations in soils and affect ecosystem health and productivity (Shortle and Smith, 1988; Cronan and Grigal, 1995). Results of these and other studies are helping to formulate needed strategies for the control and mitigation of atmospheric deposition and the protection of terrestrial and aquatic ecosystems. However, a watershed ecosystem analysis of atmospheric deposition is long-term because of temporal variations in weather and wet and dry deposition (Megahan and Hornbeck, 2000). It is necessary that some parameters be measured for decades to describe their variability. Therefore, historical data on site-specific nutrient concentrations in stream water—some spanning 30 years or more—continue to be valuable in studying the long-term trends related to natural plant succession and atmospheric deposition.

3.2.4.2. Introduced Chemicals

Pesticides, fertilizer, and other chemicals introduced on watershed lands can affect water quality and aquatic ecosystems (Norris et al., 1991). While their direct toxic effects are a major concern, these chemicals can have indirect detrimental effects on aquatic ecosystems at concentrations considerably lower than those that cause mortality. As a consequence, the effects of these chemicals must be evaluated on the basis of:

- Site-specific changes caused by the chemicals on the aquatic ecosystem of concern;
- Subsequent changes in other (often downstream) communities of aquatic organisms;
- Alterations of terrestrial ecosystems that influence the aquatic ecosystems; and
- Effects of the chemicals on patterns of recovery on watersheds that have already been altered by tree cutting or fire.

A stream is generally 5-to-10 times less likely to be affected by introduced chemicals when:

- It is not flowing through treated areas;
- Protective, riparian buffer strips have been established along the streams; and
- Attention is given to preventing drift and direct application of chemicals to the stream (Norris et al., 1991; Megahan and Hornbeck, 2000; Verry et al., 2000).

3.2.4.3. Water Temperature

Water temperature is a water-quality constituent that can affect fish production and recreational use of lakes and streams. For example, individual fish species generally have unique temperature ranges in which they thrive. Therefore, water temperature outside this range is usually detrimental to their well-being. A principle source of heat to a stream is solar energy on the water's surface (Brown, 1980; Brooks et al., 1997). Retaining the shade from overhanging riparian vegetation is an important factor in regulating the stream temperature. Coniferous trees usually provide greater shade than deciduous vegetation and can better regulate stream temperature.

3.2.5. Other Natural Resources and the Multiple-Use Concept

Vegetative manipulations on watershed lands influence the quantity, quality, and timing of streamflow regimes and affect other natural resources, often in a beneficial way. Results from many watershed experiments have shown that vegetation can often be managed to increase water yields, while providing timber, forage, wildlife, and amenity values in some optimal combination at a sustainable level (Baker and Ffolliott, 2000; Beschta, 2000; Brooks et al., 1997, 2000; Swank and Tilley, 2000; Verry et al, 2000). These findings are not surprising because many vegetative treatments imposed for their water-yield improvement possibilities are common in principle and application to management programs implemented to benefit other natural resources. However, the initial magnitude and sustainability of these site-specific changes in natural resource products and uses depend on the:

- Severity of the vegetative treatments;
- Region where they are imposed; and
- Land-use policies and management philosophy of the responsible agency.

The holistic management of natural resources on watershed lands to produce more than one product or amenity value reflects the multiple-use concept of management. In most discussions of watershed management, ecosystem-based, multiple-use management is cited as a guiding principle. The multiple-use approach to management takes advantage of the relationships among natural resources so that by manipulating one or a few of the resources, additional benefits are derived from the related resources (Brooks et al., 1992, 1997; Satterland and Adams, 1992; Jensen and Bourgeron, 1994; Kaufmann et al., 1994). Natural resources are more efficiently used by recognizing multiple-use potentials and incorporating them into comprehensive, integrated management planning. From a watershed management perspective, the ecosystem-based, multiple-use approach orients planners, managers, and decision-makers away from viewing watershed management as only water management. Planning proceeds with a sustainable approach to achieve upstream and downstream objectives. The emphasis on effectively integrating multiple-use concepts into watershed management will become increasingly important in the future as rural and urban developments extend onto pristine watershed lands.

3.3. EMERGING TOOLS AND TECHNOLOGIES

Planning and implementing successful watershed management practices depends on the availability of high-quality, readily-accessible data (Guertin et al., 2000). Over the past decade or so, many significant advances have been made in capturing, storing, and efficiently using this data (Figure 3.1). Emerging tools and technologies hold great promise for improving the understanding of ecosystem processes and revolutionizing watershed research and management. The tools and technologies include:

- Improved remote-sensing platforms;
- Global positioning systems (GPS);
- GIS; and
- The Internet.

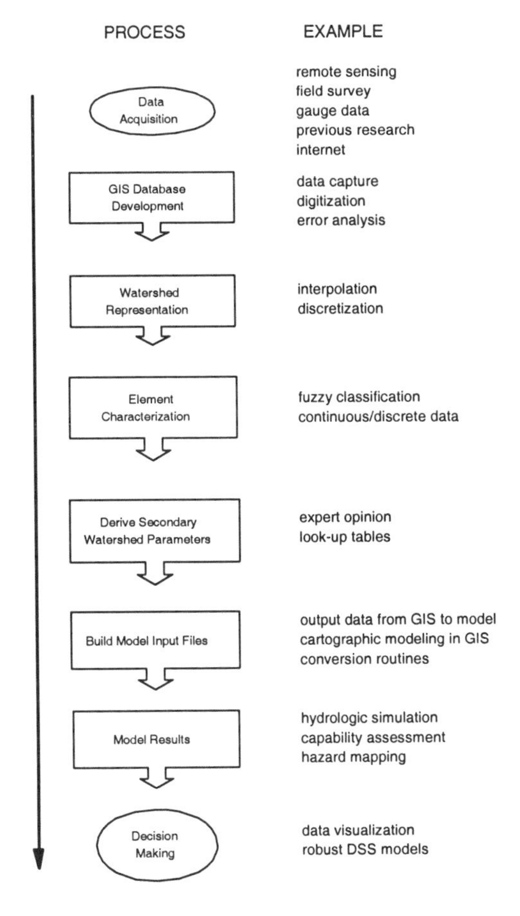

Figure 3.1. Incorporating emerging tools and technologies into watershed management practices (adapted from Guertin et al., 2000).

A diversity of decision-support systems facilitate the use of the data obtained by applying these tools and technologies in planning, implementing, and evaluating watershed management practices.

3.3.1. Data Acquisition

Ecosystem processes and land-use practices on watershed lands are generally spatially distributed and, therefore, surface characteristics of the landscapes largely determine the responses of land, water, and other natural resources to change. Consequently, assessment, modeling, and simulation techniques must account for the spatial variability of soil, geographic, topographic, vegetation, and management characteristics (Beven and Moore, 1992; Sample, 1994; Lane et al., 1997; Guertin et al., 2000). Determining the boundaries of these and other characteristics within watersheds is critical. However, traditional mapping techniques that rely on surface travel and surveying equipment are tedious, locally intensive, and relatively inaccurate. Advances in the spatial characterization of landscape features by improved remote-sensing platforms and GPS allow for rapid and precise mapping and assessment of spatially distributed surface characteristics.

3.3.1.1. Remote-sensing Platforms

Remote-sensing information about an object is obtained without physical contact (Schott, 1997). Remote-sensing technology is expanding rapidly along with advances in computing and engineering technologies (Sample, 1994; Caylor, 2000; Guertin et al., 2000). While significant advances have been made in all phases of remote sensing, the focus has often been placed on the electromagnetic (EM) spectrum. Two types of EM sensing that are used to study landscape features on watershed lands are optical remote sensing, which focuses on the short wavelength from the ultraviolet to the long-wave infrared spectra, and radar, which uses the microwave (long-wave) portion of the EM spectrum (Box 3.2). Many types of imaging, including vision, ground- and aerial-based photography, satellite observations, radar, and sonar, are widely applied in earth-science observation, modeling, and the management of land, water, and other natural resources.

3.3.1.2. Global Positioning Systems (GPS)

GPS satellites, operated by the U.S. Department of Defense, provide continuous coverage of the earth to any point, with a minimum of 4 visible satellites (Twigg, 1998; Guertin et al., 2000). These satellites broadcast 2 radio signals that carry navigation codes and messages and are received by ground sensors. Codes and messages received by the sensors are used to calculate distances among the satellites, and geometric algorithms are used to determine the precise position of the receiver. GPS satellites offer opportunities to improve on traditional field-survey techniques for watershed management. For example, a GPS satellite can be used to identify the location of stream-channel cross-sections, flumes and weirs, weather stations and remote precipitation gages, observational wells, and soil and vegetation study plots. However, GPS satellites are limited in their ability to fully characterize the spatially distributed information of a watershed and are therefore, most useful for boundary and point surveys.

3.3.2. Spatial Analysis and Modeling

An unparalleled evolution of computing technologies has occurred in recent years to the benefit of watershed managers. GIS are examples of this accelerating development. GIS have the power to collect, store, analyze, and update georeferenced information. They can be programmed to process and display this spatial information on demand in a format that meets the users' informational needs (Goodchild et al., 1993; Sample, 1994; Guertin et al., 2000). GIS can also be linked to computer simulation techniques and decision-support systems (see below) to enhance their capabilities for spatial analysis and modeling.

3.3.2.1. Analysis

GIS algorithms have been developed to delineate watershed boundaries (Band, 1986; Jensen and Dominque, 1988) and compute hillslope-flow paths (Quinn et al., 1991). GIS-based tools have also been created to extract watershed-based information from digital elevation models for watershed characterization and model parameterization (Eash 1994; Miller et al., 1999). For example, Hutchinson (1989) developed a method of gridding elevations that automatically removes spurious pits and incorporates a drainage-enforcement algorithm into the procedure to maintain fidelity with a watershed's drainage network. Interpolation routines have been developed to create spatially distributed coverage of point observations on a watershed. GIS can provide the required input data for geostatistical techniques that interpret rainfall, soil information, and contaminants (Burrough and McDonnell, 1998).

Box 3.2 Remote-sensing Platforms: Watershed Management Applications

Optical remote-sensing instruments are passive devices that record electromagnetic (EM) waves emitted from the earth's surface. Different combinations of landscape features interact with the EM spectrum in different ways. Therefore, interpreting the signals emitted for land-surface characterizations is possible. Many remote-sensing platforms are operational, others have completed their usefulness, and still others are presently in design or early production phases. The array of sensors limited early satellite platforms deployed and, consequently, they targeted small windows in the EM spectrum for only specific applications (Sample, 1994; Guertin et al., 2000).

With recent improvements in design, engineering materials, and computing power, multi-spectral platforms are now available to sense large portions of the EM spectrum, which enhance landscape classifications. Optical-remote sensing has advantages over traditional land-surface studies, which are often limited by sample size and point-based estimates of the parameters. Optical-remote sensing provides a unique and useful perspective of the earth's surface; it is also suitable for large-scale investigations of surface patterns. However, there are constraints to the use of optical-remote sensing, including operational limitations to daylight hours and the ability of clouds and smoke to mask the surface signals from the sensors. Since geologic materials, soils, vegetation, and human structures affect the EM signal in different ways, algorithms have been developed to interpret landscape features for site-specific applications, such as vegetation classification and soil analysis (Sample, 1994; Lachowski et al., 1998). Advanced image-processing tools and techniques can classify landscapes into regions upon which specific categorization-algorithms are imposed.

Wavelengths in the microwave spectrum are orders of magnitude larger than those sensed in the optical range. Radio Detection and Ranging (radar) interpret these longer wavelengths to allow for inferences about the surface properties for landscape classification. Most radar applications are active systems where a satellite or aircraft emits a microwave signal toward the object of interest and records the signal upon its return (Guertin et al., 2000). Synthetic aperture radar (SAR) is emerging as a tool that facilitates detailed mapping of surface characteristics through processing of radar signals such that the azimuth resolution is improved in direct proportion to the system's aperture-size (Henderson and Lewis, 1998; Dobson, 2000). SAR has the potential for applications in watershed management since it can provide high- resolution images, is unaffected by atmospheric conditions, is an active system, and the return signal is highly affected by the imaged target. The SAR signal can also be polarized and is coherent, providing both amplitude and phase as a target function.

Inteferometric SAR, where a target is sensed multiple times from different orientations, is used to prepare highly detailed topographic maps. Inteferometric SAR instruments are also used to detect land-surface changes, flood events, and tree-harvesting activities (Guertin et al., 2000). This technology has the potential to provide information needed to monitor and evaluate large-scale watershed management programs.

3.3.2.2. Modeling

Most watershed simulation models and other analysis techniques will use GIS to some extent in the future. GIS represent watersheds through interpolation of rainfall gage records and other point-based data and through modeling subwatershed or response-unit delineations.(Goodchild et al., 1993; Lane et al., 1997; Brown, 2000; Guertin et al., 2000; Roller and Bergen, 2000). Once a watershed has been stratified into the required modeling elements through these interpolations, the required modeling inputs characterize each element, and these inputs are entered into the model. The model is incorporated entirely into a GIS using cartographic modeling techniques in some applications (Tomlin, 1990). The products of this approach are usually new GIS coverages containing the model results. Delineation of land capacity or suitability (Sheng et al., 1997), landslide-hazard mapping (Carra et al., 1991; Montgomery and Dietrich, 1994), and locating erosion hazards (Warren, 1989) are examples of this type of analysis. Alternatively, the model is external to the GIS, but it uses the GIS output data for its parameterization. Examples of this class include the Agricultural Nonpoint Source model of pollution on a watershed (Young et al., 1989) and the Water Erosion Prediction Project model used to predict erosion (Savabi et al., 1995).

3.3.2.3. Error Assessment

While applications of GIS are rapidly increasing, watershed researchers and managers must recognize the issues surrounding spatially distributed errors and model behavior (Bolstad and Smith, 1992; Longley et al., 1998; Guertin et al., 2000). Errors can occur in field measurements, the original map, and digitizing efforts. These errors can be introduced at almost any step in integrating GIS-based processes into spatial analysis and models. Some of these errors are cumulative, causing the spatial analysis or modeling effort to be incorrect or nonrepresentative. The effects of these errors on decision making continue to be studied through sensitivity analyses.

3.3.3. Internet Applications

The Internet provides researchers, managers, and decision-makers with revolutionary access for information transfer. The Internet mirrors the diversity of subject-matter information that is available worldwide (Lawrence and Giles, 1998; Johnson, A., 2000; Huebner et al., 2000). Internet communications frequently used by watershed managers to exchange information include:

- Electronic mail (email) messages,
- Bulletin boards, and
- The World Wide Web.

3.3.3.1. Electronic Messages

The most widespread form of Internet communication is email (Johnson, A., 2000). Email is a useful, easy-to-use tool for delivering content and receiving feedback about watershed-related issues. Email messages are generally brief and focused. A subscriber maillist—a collection of email addresses—allows users to send messages of common interest to many recipients. When email messages become unwieldy, bulletin boards give users an effective communication method.

3.3.3.2. Bulletin Boards

Bulletin boards are established so that a conversation on any topic is maintained and participants may access the communication anytime to read or respond to discussion comments (Johnson, A., 2000). Bulletin boards, also called message boards, conferencing systems, and asynchronous discussion forums are effective when archived conversations need to be accessed for future reference and consideration. By contrast, chat conversations occur in real-time and are not archived for future reference.

3.3.3.3. World Wide Web

The World Wide Web is a system of Internet servers that support specially formatted documents with links to other documents having graphics and audio and video files. Users can find publications, data sets, images, software, and bibliographies about watershed management on many Website formats (Box 3.3). Subject guides and search engines have been developed to help watershed managers effectively gather and distribute this information (Baker et al., 2000a, 2000b; Johnson, A., 2000).

Subject guides are hierarchically organized indexes of subject-matter categories that allow the user to browse through lists of Websites by subject-matter for information (Huebner et al., 2000). General subject guides can be suitable for exploring informational resources about broad topics, such as arts and humanities. If the desired information is more specific, for example, the streamflow records for watersheds in a particular region, specialized search

Box 3.3 Beaver Creek Website

Information from the Beaver Creek watershed project in north-central Arizona has been incorporated into a Website (http://ag.arizona.edu/OALS/watershed) for ease of communication and information sharing (Baker et al., 2000c). Data on this Website include precipitation, air temperature, and relative humidity; streamflow amounts and distribution; sediment yields and water-quality characteristics; and information on forage and timber production and wildlife populations and habitat qualities. The Website contains links to the various categories. Search lists of the types of information contained on the site are available to the user. Drop-down lists are also available for easy access to the data sets for selected watersheds and particular years. An overview provides a narrative on the Beaver Creek watershed project, which includes a site description and history and highlights of research findings. A publication link contains nearly 700 annotated citations of publications and reports on the Beaver Creek project.

The information on the Beaver Creek Website provides information to watershed researchers, managers, and decision-makers to help them resolve current and future land stewardship issues. Although the data sets are public-domain information, they are only minimally useful if knowledge of their existence and physical accessibility limits their access. Accessing these data bases through the Web allows interested individuals to download them into software packages and models that did not exist when the data were collected. Data sets from Beaver Creek find applications in many arid and semiarid regions of the world and provide long-term, readily accessible resource data for new analysis techniques and modeling applications.

engines are helpful. Because there are thousands of specialized subject guides available, clearinghouse sites are often available to help the user efficiently access them.

Search engines are developed by building an index from existing Websites and then giving the user the ability to query that index (Huebner et al., 2000; Johnson, A., 2000). To build a desired index, search engines deploy software robots that automatically index the contents of a Website. The robot indexes the Web pages linked to the first page and then moves through cascading myriads of linked pages. Because of the automation used in their development, search engines can index a larger portion of the Web than subject guides. A larger index means that more pages relating to a narrowly focused topic will be found and delivered to the watershed manager. However, because search engines index many Websites, a large portion of the pages can lack relevancy; this is especially true if the query is too broad.

Another way of obtaining information is to attach Websites addresses to communications with colleagues. Information about Web pages is also in newsletters, bulletin boards, and mailing lists. Websites have become the universe of network-accessible information for watershed management.

3.3.4. Decision-Support Systems

Watershed managers are faced with the increasingly difficult task of making appropriate, responsible, and acceptable decisions on the use and management of land, water, and other natural resources. The complexity of the questions asked by the public, the extensive information available, and the dwindling amounts of land, water, and other resources are why better decision-making processes have become necessary. Fortunately, a diversity of powerful decision-support systems have been recently developed to help managers make better decisions about watershed management (Lane and Nichols, 2000). Two common formulations of these systems are linear programming and multiple-criteria decision making (MCDM).

3.3.4.1. Linear Programming

Linear programming is the basis for relatively simple decision-support systems. For example, a watershed manager might wish to reduce the costs of a management practice represented by a linear-objective functioning and a set of linear constraints—the objective function and all of the constraints are strictly linear over the range of permissible values. However, this basic assumption of linearly is often inappropriate or unsuitable for complex watershed management problems.

3.3.4.2. Multiple-criteria Decision Making

Decision-making problems more commonly confronted by watershed managers usually involve several objectives. For example, reducing the costs of a management practice while simultaneously optimizing the total benefits obtained from forage production, wildlife-habitat protection, and soil and water conservation practices. A set of such objective functions would be subject to several linear and nonlinear constraints. Complex but realistic problems are appropriately analyzed with MCDM methods (Kent and Davis, 1988; Yakowitz and Szidarovszky, 1993; Yakowitz et al., 1993; Anderson et al., 1994; El-Swaify and Yakowitz, 1998). MCDM techniques determine the preferred solutions to management-related problems in which the discrete alternatives are evaluated against acceptance criteria (factors) ranging from quantitative to qualitative (Box 3.4).

Implementation of MCDM methods to solve watershed management problems is more responsive to the public when there is participation in the process to ensure that the decision making is equitable and constructive. Non-interventionist approaches to participatory MCDM systems have emerged to achieve this purpose (Rickson et al., 1995; Lawrence et al., 2000). These processes empower stakeholders to explore all feasible options. These participatory processes integrate stakeholder perspectives and ensure that all participants view the problem similarly. This is particularly important when the problems involve a large amount of technical information. The opportunity to resolve problems and attain a consensus improves when stakeholders are involved early in the decision-making process.

Box 3.4 Applications of Multiple-criteria Decision Making (MCDM) in Watershed Management

Applications of MCDM methods in decision-support systems continue to be used to help solve complex watershed management problems. Eskandari et al. (1994, 1995a, 1995b, 1995c) and de Steiguer (2000) studied the problems of implementing ecosystem-based, multiple-use management in water-yield improvement programs on upland watersheds. Issues of uncertainty in the decision-making process due to the inherent variability exhibited in hydrologic and ecosystem functioning was also considered. Yakowitz et al. (1992a, 1992b) applied MCDM methods to evaluate the possible effects of alternative watershed management practices on nonpoint source pollution and water quality and in selecting the best course of action.

3.4. LOCALLY LED WATERSHED MANAGEMENT INITIATIVES

Sociocultural considerations have been increasingly incorporated into the planning and implementation of watershed management programs—a trend that is likely to continue. To achieve integration of sociocultural considerations into these programs, the goals and objectives of public management and regulatory agencies and the diverse interests of the public are incorporated using a variety of watershed partnerships, councils, and corporations and other locally-led watershed management initiatives.

More than 1,500 locally-led watershed management initiatives have been established—most within the 1990s—in the United States (Lant, 1999). Similar initiatives have also been formed in Brazil, Australia, and several other countries. Interactions of the social, political, and economic forces of land stewardship, with the technical aspects of watershed management, are effectively fostered through the activities of these organizations.

3.4.1. Watershed Partnerships

The basic idea of watershed partnerships is that all members have equal decision-making power (Toupal and Johnson, 1998; Lant, 1999; Endebrock, 2000; Garcia, 2000; Johnson, M.D.,

2000). Successful watershed partnerships begin with leadership provided by local stakeholders and representatives of federal, state, or other public management or regulatory agencies. Active participation in watershed partnerships is recruited from a variety of stakeholders representing a diversity of interests.

Leadership and participants of a watershed partnership and other locally-led initiatives must be clear and consistent with their communications. Adaptability and flexibility must be encouraged, common goals must be established, and trust must be built. Consensus-building decision making is necessary throughout the relationship. When necessary, financial support must be identified and secured. Development of effective watershed partnerships is expected to continue into the future (Box 3.5). Successful watershed partnerships are an effective and efficient way to achieve conservation goals and effective, future land stewardship.

3.4.2. Watershed Councils

Watershed councils are also locally-oriented and function as nonprofit advisory, educational, or advocacy organizations that encourage river-basin protection, conservation, and sustainability of natural resources. Some watershed councils are forums for exchanging ideas, views, concerns, and recommendations, while others are actively responsible for river-basin management. Similar to watershed partnerships, membership in watershed councils is usually open to public agency personnel and private-sector stakeholders, with technical and sociocultural interests in the river basin and its tributary watersheds. Communication among council members is largely through open meetings, newsletters, and information collected and distributed on the Internet. Organization of watershed councils has accelerated greatly in the past 10 years, largely because of increasing public concern about and awareness of ecology, environmental quality, and the sociocultural perspectives of watershed management. Watershed councils are currently in every region of the United States.

Watershed-council mission statements reflect the concern of watershed managers and stakeholders in a diversity of local issues. For example:

- The Saugus River Watershed Council of Massachusetts is concerned with the health and beauty of the Saugus River and its tributary watersheds.
- The Watershed Agricultural Council promotes environmentally and economically sound agricultural and forestry practices, while protecting the New York City water supply for more than 9 million consumers.
- The Connecticut River Watershed Council is the principal advocacy group for protecting and conserving the Connecticut River, and linking the council members to other organizations and agencies.
- The Chetco River Watershed Council manages the Chetco River Watershed, which runs from western Oregon to the Pacific Ocean.

The Watershed Management Council is a nonprofit, nationwide educational organization dedicated to watershed management. The Council publishes a newsletter on timely themes and events, such as conferences on interdisciplinary collaboration, cumulative watershed effects, riparian systems, watershed restoration, and water-quality monitoring.

Box 3.5 Development of Effective Watershed Partnerships

The Natural Resources Conservation Service of the U.S. Department of Agriculture has taken a key leadership role at the federal level in developing watershed partnerships in the United States. In this leadership capacity, the Natural Resources Conservation Service has identified 4 factors that have helped to develop effective partnerships. A loss of trust in natural resource management and enforcement of environmental legislation and programs by public land management and regulatory agencies, the first factor, is attributed to multiple causes (Toupal and Johnson, 1998; Johnson, M.D., 2000). These causes include:

- lack of a clear environmental agenda that extends beyond a single presidential administration;
- fragmentation of compliance responsibilities among multiple agencies; and
- lack of funding necessary to implement legal restrictions and programs (Lazarus, 1991: Brooks et al., 1994; Quinn et al., 1995).

This lack of trust has also caused nongovernmental organizations, private interest groups, and local citizens to assume an adversarial role concerning the actions of public land management and regulatory agencies.

A second factor leading to the development of effective watershed partnerships is the increased access to technical information by the public (Johnson, M.D., 2000). Every person that has Internet access can receive large amounts of information on nearly any subject. The Internet is becoming the preferred method of distributing information for public agencies and the private sector (Tapscott, 1999; Wolinsky, 1999). Through the Internet, private-sector partners can provide immediate feedback to public land management and regulatory agencies in the planning process. These partners can also raise issues and concerns before the planning process proceeds on incorrect assumptions. Increased availability of technical information has also raised public awareness about the activities of land management and regulatory agencies. Many privately financed environmental awareness organizations maintain a continuous watch over public agencies by monitoring their planning and environmental compliance activities (Toupal and Johnson, 1998). These organizations use multimedia approaches to highlight the perceived mistakes or failures of public land agencies to comply with environmental laws and to bring this information to the public's attention in a short time.

A third factor that has caused effective watershed partnerships is the increased focus placed on non-commodity aspects of natural resources, such as the recreational, landscape beauty, and indigenous beliefs about land, water, and other natural resources (Brunson, 1996; Griffin, 1999). While determining the monetary value of the non-commodity aspects of natural resources is difficult, many people think that they are vital to watershed management and should be considered in the planning process. Most of the non-commodities are also difficult to address in the planning process without recognition of the relationship between individuals and planners (Johnson, M.D., 2000). One way to cultivate these relationships is through development and use of shared-power partnerships rather than the more traditional top-down approach to the planning process (Austin, 1998; Nazarea et al., 1998).

A fourth factor that has led to the development of effective watershed partnerships is an increasing demand by the public to incorporate the multiple-use concept into management planning actively and meaningfully. No longer is natural resource management driven only by the interests of a single group of stakeholders or once economic concern (Cleary, 1988; Kaufmann et al., 1994; Brunson, 1996). This is crucial to the success of conservation partnerships (Toupal and Johnson, 1998). Appropriate incorporation of multiple interests and concerns in land, water, and other natural resource management is the goal of most watershed partnerships.

3.4.3. Watershed Corporations

Watershed corporations are usually nonprofit organizations consisting of members from communities within a river basin, from the responsible public management and regulatory agencies, and elected officials. One example of a watershed corporation is the Catskill Watershed Corporation, which oversees management of the drainage basins that flow into key reservoirs from the Catskill Watershed to sustain high-quality drinking water for the residents of New York City (Garcia, 2000). The Catskill Watershed Corporation recognized the need to consider social, economic, political, and technical issues as a set of geographic units defined by political boundaries and another set of geographic units delineated by watershed boundaries. Members of this watershed corporation acknowledge the potential for conflict between New York City's growth and development and preservation and restoration of the quality of water flowing into the city's reservoirs. Because the corporation anticipates the need to accommodate these competing values, it is well-structured to facilitate informed decision making by all parties.

3.5. SUMMARY

A perspective of issues to be confronted in land stewardship and the contributions of watershed management to future land stewardship is enhanced by taking a retrospective view of the history of watershed management, the lessons learned, the emerging tools and technologies, and locally-led initiatives. Watershed management has evolved from an understanding of the importance of the hydrologic cycle to a focus on water resource management to a holistic and integrated approach of managing the biological, physical, and social elements on a landscape delineated by watershed boundaries. Many studies and extensive management experiences have shown that planned and unplanned vegetative manipulations can alter streamflow regimes, erosion and sedimentation processes, water-quality characteristics, and the flow of other natural resources from watershed lands.

Information obtained and lessons learned from these studies and experiences have been incorporated into watershed management practices, projects, and programs to furnish a foundation for future land stewardship. Planning and implementing these practices, projects, and programs depends on the availability of high-quality and accessible information. Emerging tools and technologies, including improved remote-sensing platforms, GPS, GIS, and the Internet, hold promise for improving the understanding of ecosystem processes and revolutionizing watershed research and management. Sociocultural considerations have been integrated into the planning and implementation of watershed management programs in recent years to strengthen these interventions. To achieve integration of sociocultural considerations, the goals and objectives of public management and regulatory agencies and diverse public interests are being integrated through a variety of watershed partnerships, watershed councils, watershed corporations, and other locally-led watershed management initiatives.

REFERENCES

Anderson, D. R., Sweeney, D. J., Williams, T. A., *An Introduction to Management Science: Quantitative Approaches to Decision Making*. St. Paul, MN: West Publishing Company, 1994.

Angino, E. E., Magnuson, L. M., Waugh, T. C. Mineralogy of suspended sediment and concentrations of Fe, Mn, Ni, Zn, Cu, and Pd in water and Fe, Mn, and Pd in suspended load of selected Kansas streams. Water Resources Research 1974; 10:1187-1191.

Austin, D. Cultural knowledge and the cognitive map. Practicing Anthropology 1998; 20:21-24.

Baker, M. B., Jr., compiler, *History of Watershed Research in the Central Arizona Highlands*. Fort Collins, CO: Rocky Mountain Research Station, USDA Forest Service, General Technical Report RMRS-GTR-29, 1999.

Baker, M. B., Ffolliott, P. F. "Contributions of Watershed Management Research to Ecosystem-Based Management in the Colorado River Basin." In *Land Stewardship in the 21st Century: The Contributions of Watershed Management*, P. F. Ffolliott, M. B. Baker, Jr., C. B. Edminster, M. C. Dillon, and K. L. Mora, tech. coords. Fort Collins, CO: Rocky Mountain Research Station, USDA Forest Service, Proceedings RMRS-P-13, 2000, pp. 117-128.

Baker, M. B., Jr., Huebner, D. P., Ffolliott, P. F. "Accessing a Personalized Bibliography with a Searchable System on the World Wide Web." In *Land Stewardship in the 21st Century: The Contributions of Watershed Management*, P. F. Ffolliott, M. B. Baker, Jr., C. B. Edminster, M. C. Dillon, and K. L. Mora, tech. coords. Fort Collins, CO: Rocky Mountain Research Station, USDA Forest Service, Proceedings RMRS-P-13, 2000a, pp. 428-430.

Baker, M. B., Jr., Huebner, D. P., Ffolliott, P. F. "An On-line Image Data Base System: Managing Image Collections." In *Land Stewardship in the 21st Century: The Contributions of Watershed Management*, P. F. Ffolliott, M. B. Baker, Jr., C. B. Edminster, M. C. Dillon, and K. L. Mora, tech. coords. Fort Collins, CO: Rocky Mountain Research Station, USDA Forest Service, Proceedings RMRS-P-13, 2000b, pp. 424-427.

Baker, M. B., Jr., Huebner, D. P., Ffolliott, P. F. "Documenting Historical Data and Accessing it on the World Wide Web." In *Land Stewardship in the 21st Century: The Contributions of Watershed Management*, P. F. Ffolliott, M. B. Baker, Jr., C. B. Edminster, M. C. Dillon, and K. L. Mora, tech. coords. Fort Collins, CO: Rocky Mountain Research Station, USDA Forest Service, Proceedings RMRS-P-13, 2000c, pp. 189-193.

Band, L. E. Topographic partition of watersheds with digital elevation models. Water Resources Research 1986; 22:15-24.

Bates, C. H., Henry, A. J. Forest and streamflow experiments at Wagon Wheel Gap, Colorado. U.S. Weather Bureau Monthly Weather Review Supplement 30, 1928.

Benda, L., Dunne, T. Stochastic forcing of sediment routing and storage in channel networks. Water Resources Research 1997; 33:2849-2863.

Beschta, R. L. "Watershed Management in the Pacific Northwest: The Historical Legacy." In *Land Stewardship in the 21st Century: The Contributions of Watershed Management*, P. F. Ffolliott, M. B. Baker, Jr., C. B. Edminster, M. C. Dillon, and K. L. Mora, tech. coords. Fort Collins, CO: Rocky Mountain Research Station, USDA Forest Service, Proceedings RMRS-P-13, 2000, pp. 109-116.

Beven, K. J., Moore, I. D., eds., *Terrain Analysis and Distributed Modeling in Hydrology*. New York: John Wiley & Sons, Inc., 1992.

Bolstad, P. V., Smith, J. L. Errors in GIS. Journal of Forestry 1992; 90(11):21-29.

Bosch, J. H., Hewlett, J. D. A review of catchment experiments to determine the effect of vegetation changes on water yield and evapotranspiration. Journal of Hydrology 1982; 55:2-23.

Brooks, K. N., Ffolliott, P. F. "Forest Hydrology." In *Forests - A Global Perspective*, S. K. Majumdar, Miller, E. W., Brenners, P. J., eds. Easton, PA: The Pennsylvania Academy of Science, 1996, pp. 26-40.

Brooks, K. N., Ffolliott, P. F., Gregersen, H. M., DeBano, L. F., *Hydrology and the Management of Watersheds*. Ames, IA: Iowa State University Press, 1997.

Brooks, K. N., Ffolliott, P. F., Gregersen, H. M., Easter, K. W., *Policies for Sustainable Development: The Role of Watershed Management*. Washington, DC: U.S. Department of State, EPAT Policy Brief 6, 1994.

Brooks, K. N., Gregersen, H. M., Ffolliott, P. F., Tejwani, K. G. "Watershed Management: A Key to Sustainability." In *Managing the World's Forests: Looking for Balance Between Conservation and Development*, N. P. Sharma, ed. Dubuque, IA: Kendall/Hunt Publishing Company, 1992, pp. 455-487.

Brown, D. G. Image and spatial analysis software tools. Journal of Forestry 2000; 98(6):53-57.

Brown, G. W., *Forestry and Water Quality*. Corvallis, OR: Oregon State University Book Stores, 1980.

Brown, T. C., Brown, C., Binkley, D. Laws and programs for controlling nonpoint source pollution in forest areas. Water Resources Bulletin 1993; 29:1-13.

Brunson, M. W. Integrating human habitat requirements into ecosystem management strategies: A case study. Natural Areas Journal 1996; 16:100-107.

Burrough, P. A., McDowell, R. A., *Principles of Geographic Information Systems*. New York: Oxford University Press, 1998.

Burroughs, E. R., Jr., King, J. G., *Reduction of Soil Erosion on Forest Roads*. Ogden, UT: Intermountain Forest and Range Experiment Station, USDA Forest Service, General Technical Report INT-264, 1989.

Campbell, R. E., Baker, M. B., Jr., Ffolliott, P. F., Larson, F. R., Avery, C. C., *Wildland fire Effects on a Ponderosa Pine Ecosystem: An Arizona Case Study*. Fort Collins, CO: Rocky Mountain Forest and Range Experiment Station, USDA Forest Service, Research Paper RM-191, 1977.

Carra, A., Cardinali, M., Detti, R., Guzzetti, F., Pasqui, V., Reichenback, P. GIS techniques and statistical models in evaluating landslide hazard. Earth Surface Processes and Landforms 1991; 16:427-445.

Caylor, J. Aerial photography in the next decade. Journal of Forestry 2000; 98(6):17-19.

Chandra, S., *Hydrology in Ancient India*. Roorke, India: National Institute of Hydrology, 1990.

Chow, V. T., ed., *Handbook of Applied Hydrology: A Compendium of Water-Resources Technology*. New York: McGraw-Hill Book Company, 1964a.

Chow, V. T., "Section 14: Runoff." In *Handbook of Applied Hydrology: A Compendium of Water-Resources Technology*, V. T. Chow, ed. New York: McGraw-Hill Book Company, 1964b, pp. 14-1-14-54.

Cleary, C. R. Coordinated resource management: A planning process that works. Journal of Soil and Water Conservation 1988; 43:138-139.

Colman, E. A., *Vegetation and Watershed Management*. New York: The Ronald Press, 1953.

Cronan, C. S., Grigal, D. F. Use of calcium/aluminum ratios as indicators of stress in forest ecosystems. Journal of Environmental Quality 1995; 24:209-226.

Cronan, C. S., Schofield, C. L. Relationships between aqueous aluminum and acidic deposition in forested watersheds of North America and northern Europe. Environmental Science and Technology 1990; 24:1100-1105.

DeBano, L. F., Neary, D. G., Ffolliott, P. F., *Fire Effects on Ecosystems*. New York: John Wiley & Sons, Inc., 1998.

de Steiguer, J. E. "Applying EXCEL Solver to a Watershed Management Goal-Programming Problem." In *Land Stewardship in the 21st Century: The Contributions of Watershed Management*, P. F. Ffolliott, M. B. Baker, Jr., C. B. Edminster, M. C. Dillon, and K. L. Mora, tech. coords. Fort Collins, CO: Rocky Mountain Research Station, USDA Forest Service, Proceedings RMRS-P-13, 2000, pp. 325-329.

Dixon, J. W. "Section 26-1: Water Resources - Part I: Planning and Development." In *Handbook of Applied Hydrology: A Compendium of Water-Resources Technology*, V. T. Chow, ed. New York: McGraw-Hill Book Company, 1964, pp. 26-1-26-29.

Dobson, M. C. Forest information from synthetic aperture radar. Journal of Forestry 2000; 98(6):41-43.

Dortignac, E. J. "Forest Water Yield Management Opportunities." In *Forest Hydrology: Proceedings of a National Science Foundation Advanced Seminar*, W. E. Sopper and H. W. Lull, eds. New York: Pergamon Press, 1967, pp. 579-592.

Dunne, T., Leopold, L. B., *Water in Environmental Planning*. San Francisco, CA: W. H. Freeman Company, 1978.

Eash, D. E. A geographic information system procedure to quantify drainage-basin characteristics. Water Resources Bulletin 1994; 30:1-8.

Eckman, K., Gregersen, H. M., Lundgren, A. L. "Watershed Management and Sustainable Development: Lessons Learned and Future Directions." In *Land Stewardship in the 21st Century: The Contributions of Watershed Management*, P. F. Ffolliott, M. B. Baker, Jr., C. B. Edminster, M. C. Dillon, and K. L. Mora, tech. coords. Fort Collins, CO: Rocky Mountain Research Station, USDA Forest Service, Proceedings RMRS-P-13, 2000, pp. 37-43.

Elliot, W. J. "Modeling Rangeland Watershed Erosion Process." In *Watershed 2000: Science and Engineering Technology for the New Millennium*, M. Flug and D. Frevert, eds. Reston, VA: American Society of Civil Engineers, 2000. [CD-ROM] Windows.

El-Swaify, S. A., Yakowitz, D. S., ed., *Multiple Objective Decision Making for Land, Water, and Environmental Management*. Boca Raton, FL: Lewis Publishers, 1998.

Endebrock, E. G. "Arizona Rural Watershed Initiative: Addressing Arizona's Rural Watershed Needs Through Regional Partnerships." In *Watershed 2000: Science and Engineering Technology for the New Millennium*, M. Flug and D. Frevert, eds. Reston, VA: American Society of Civil Engineers, 2000. [CD-ROM] Windows.

Eskandari, A., Ffolliott, P., Szidarovszky. Multicriterion decision-making for sustainable watershed resources management. PUMA 1994; 5:379-390.

Eskandari, A., Ffolliott, P., Szidarovszky. "Decision support system in watershed management." In *Watershed Management: Planning for the 21st Century*, T. J. Ward, ed. New York: American Society of Civil Engineers, 1995a, pp. 208-217.

Eskandari, A., Ffolliott, P., Szidarovszky. Uncertainty and method choice in discrete multiobjective programming problems. Applied Mathematics and Computation 1995b; 69:335-351.

Eskandari, A., Ffolliott, P., Szidarovszky. Uncertainty in multicriterion watershed management problems. Technology: Journal of the Franklin Institute 1995c; 332A:199-207.

Ffolliott, P. F., Brooks, K. N. "Process Studies in Forest Hydrology: A Worldwide Review." In *Surface-Water Hydrology*, V. P. Sing, Kuman, B., eds. Dordrecht, The Netherlands: Kluwer Academic Publishers, 1996, pp. 1-18.

Ffolliott, P. F.; Baker, M. B. Jr.; Edminster, C. B.; Dillon, M. C.; Mora, K. L., tech. coords., *Land Stewardship in the 21st Century: The Contributions of Watershed Management.* Fort Collins, CO: Rocky Mountain Research Station, USDA Forest Service, Proceedings RMRS-P-13, 2000.

Fisher, S. G., Minckley, W. L. Chemical characteristics of a desert stream in flash flood. Journal of Arid Environments 1978; 16:25-33.

Frasier, G. W., Holland, K. A. "Forty Years of Rangeland Hydrology Research: Are We Making Progress." In *Watershed 2000: Science and Engineering Technology for the New Millennium*, M. Flug and D. Frevert, eds. Reston, VA: American Society of Civil Engineers, 2000. [CD-ROM] Windows.

Garcia, M. W. "New York City's Approach to Watershed Management." In *Watershed 2000: Science and Engineering Technology for the New Millennium*, M. Flug and D. Frevert, eds. Reston, VA: American Society of Civil Engineers, 2000. [CD-ROM] Windows.

Gifford, G. F., Busby, F. E. Loss of particulate organic materials from semiarid watersheds as a result of extreme hydrologic events. Water Resources Research 1973; 9:1443-1449.

Goodchhild, M. R., Parks, B. O., Steyaert, L. T., eds., *Environmental Monitoring with GIS.* New York: Oxford University Press, 1993.

Gosz, J. R., White, C. S., Ffolliott, P. F. Nutrient and heavy metal transport capabilities of sediment in the southwestern United States. Water Resources Bulletin 1980; 16:927-933.

Griffin, C. B. Watershed Councils: An emerging form of public participation in natural resources management. Journal of the American Water Resources Association 1999; 35:505-518.

Guertin, D. P., Miller, S. N., Goodrich, D. G. "Emerging Tools and Technologies in Watershed Management." In *Land Stewardship in the 21st Century: The Contributions of Watershed Management*, P. F. Ffolliott, M. B. Baker, Jr., C. B. Edminster, M. C. Dillon, and K. L. Mora, tech. coords. Fort Collins, CO: Rocky Mountain Research Station, USDA Forest Service, Proceedings RMRS-P-13, 2000, pp. 194-204.

Harr, R. D. Effects of clearcutting on rain-on-snow runoff in western Oregon; A new look at old studies. Water Resources Bulletin 1986; 22:1195-1100.

Heede, B. H., Harvey, M. D., Larid, J. R. Sediment delivery linkages in a chaparral watershed following a wildland fire. Environmental Management 1988; 12:349-358.

Henderson, F. M., Lewis, A. J., eds., *Principles and Applications of Imaging Radar.* New York: John Wiley & Sons, Inc., 1998.

Hewlett, J. D. Will water demand dominate forest management in the East? Proceedings of Annual Meeting of the Society of American Foresters. Washington, DC: Society of American Foresters, 1967, 154-159.

Hibbert, A. R. "Forest Treatment Effects on Water Yield." In *Forest Hydrology: Proceedings of a National Science Foundation Advanced Seminar*, W. E. Sopper and H. W. Lull, eds. New York: Pergamon Press, 1967, pp. 527-543.

Hibbert, A. R., *Managing Vegetation to Increase Flow in the Colorado River.* Fort Collins, CO: Rocky Mountain Forest and Range Experiment Station, USDA Forest Service, General Technical Report RM-66, 1979.

Hornbeck, J. W., Swank, W. T. Watershed ecosystem analysis as a basis for multiple-use management of eastern forests. Ecological Applications 1992; 2:238-247.

Hornbeck, J. W., Martin, C. W., Egar, C. Summary of water yield experiments at Hubbard Brook Experimental Forest, New Hampshire. Canadian Journal of Forest Research 1997; 27:2043-2052.

Hornbeck, J. W., Adams, M. B., Corbett, E. S., Verry, E. S., Lynch, J. A. Long-term impacts of forest treatments on water yield: A summary for southeastern USA. Journal of Hydrology 1993; 150:323-344.

Huebner, D. P., Baker, M. B., Jr., Ffolliott, P. F. "Increasing Efficiency of Information Dissemination and Collection Through the World Wide Web." In *Land Stewardship in the 21st Century: The Contributions of Watershed Management*, P. F. Ffolliott, M. B. Baker, Jr., C. B. Edminster, M. C. Dillon, and K. L. Mora, tech. coords. Fort Collins, CO: Rocky Mountain Research Station, USDA Forest Service, Proceedings RMRS-P-13, 2000, pp. 420-423.

Hutchinson, M. F. A new procedure for gridding elevation and stream line data with automatic removal of spurious pits. Journal of Hydrology 1989; 106:211-232.

Jensen, A. F., Domingue, J. O. Extracting topographic structure from digital elevation data for geographic information system analysis. Photogrammetric Engineer and Remote Sensing 1988; 54:1593-1600.

Jensen, M. E., Bourgeron, tech. eds., *Ecosystem Management: Principles and Applications*. Portland, OR: Pacific Northwest Forest and Range Experiment Station, USDA Forest Service, General Technical Report PNW-GTR-318, 1994.

Johnson, A. "An Effective Method of Utilizing the Internet to Enhance Watershed Management." In *Watershed 2000: Science and Engineering Technology for the New Millennium*, M. Flug and D. Frevert, eds. Reston, VA: American Society of Civil Engineers, 2000. [CD-ROM] Windows.

Johnson, M. D., "A Sociocultural Perspective on the Development of U.S. Natural Resource Partnerships in the 20th century." In *Land Stewardship in the 21st Century: The Contributions of Watershed Management*, P. F. Ffolliott, M. B. Baker, Jr., C. B. Edminster, M. C. Dillon, and K. L. Mora, tech. coords. Fort Collins, CO: Rocky Mountain Research Station, USDA Forest Service, Proceedings RMRS-P-13, 2000, pp. 205-212.

Johnson, R. S., Tew, R. K., Doty, R. D., *Soil Moisture Depletion and Estimated Evapotranspiration on Utah Mountain Watersheds*. Ogden, UT: Intermountain Forest and Range Experiment Station, USDA Forest Service, Research Paper INT-67, 1969.

Kauffmann, M. R., Graham, R. T., Boyce, D. A., Jr., Moir, W. H., Perry, L., Reynolds, R. T., Bassett, R. L., Mehlhop, P., Edminster, C. B., Block, W. M., Corn, P. S., *An Ecological Basis for Ecosystem Management*. Fort Collins, CO: Rocky Mountain Forest and Range Experiment Station, USDA Forest Service, General Technical Report RM-246, 1994.

Kent, B. M., Davis, L. S., tech. coords., *The 1988 Symposium on Systems Analysis in Forest Resources*. Fort Collins, CO: Rocky Mountain Forest and Range Experiment Station, USDA Forest Service, General Technical Report RM-61, 1988.

Kittredge, J., *Forest Influences*. New York: McGraw-Hill Book Company, 1948.

Lachowski, H., Fish, H., Brohman, R. "Riparian Area Management - The Role of Remote Sensing and Geographic Information Systems." In *Rangeland Management and Water Resources: Proceedings of the American Water Resources Association Specialty Conference*, D. F. Potts, ed. Herndon, VA: American Water Resources Association, TPS-98-1, 1998, pp. 45-54.

Lane, L. J., Nichols, M. H. "America's Watersheds: Technical Basis for New Strategies." In *Watershed 2000: Science and Engineering Technology for the New Millennium*, M. Flug and D. Frevert, eds. Reston, VA: American Society of Civil Engineers, 2000. [CD-ROM] Windows.

Lane, S. N., Richards, K. S., Chandler, J. H., eds., *Land Monitoring, Modeling and Analysis*. New York: John Wiley & Sons, Inc., 1997.

Lant, C. L. Introduction - human dimensions of watershed management. Journal of the American Water Resources Association 1999; 35: 483-486.

Lawrence, P., Shaw, R., Lane, L., Eisner, R. "Participatory Multiple Objective Decision Making Processes: Emerging Approaches with New Challenges." In *Watershed 2000: Science and Engineering Technology for the New Millennium*, M. Flug and D. Frevert, eds. Reston, VA: American Society of Civil Engineers, 2000. [CD-ROM] Windows.

Lawrence, S., Giles, C. L. Searching the World Wide Web. Science 1998; 280:98.

Lazarus, R. J. The tragedy of distrust in the implementation of federal environmental law. Law and Contemporary Problems 1991; 54:311.

Leaf, C. A., *Watershed Management in the Rocky Mountain Subalpine Zone: The Status of Our Knowledge*. Fort Collins, CO: Rocky Mountain Forest and Range Experiment Station, USDA Forest Service, Research Paper RM-137, 1975.

Longley, P. A., Goodchild, M., Maguire, D., Rhind, D. W., *Geographic Information Systems: Principles, Techniques, Applications and Management*. New York: John Wiley & Sons, Inc., 1998.

Lull, H. W., Reinhart, K. G., *Forest and Foods in the Eastern United States*. Upper Darby, PA: Northeastern Forest Experiment Station, USDA Forest Service, Research Paper NE-226, 1972.

Martin, C. W. Soil disturbance by logging in New England - Review and management recommendations. Northern Journal of Applied Forestry 1988; 5:30-34.

Martin, C. W., Hornbeck, J. W. Logging in New England need not cause sedimentation in streams. Northern Journal of Applied Forestry 1994; 11:17-23.

Megahan, W. F., Hornbeck, J. "Lessons Learned in Watershed Management: A Retrospective View." In *Land stewardship in the 21st Century: The Contributions of Watershed Management*, P. F. Ffolliott, M. B. Baker, Jr., C. B. Edminster, M. C. Dillon, and K. L. Mora, tech. coords. Fort Collins, CO: Rocky Mountain Research Station, USDA Forest Service, Proceedings RMRS-P-13, 2000, pp. 177-188.

Megahan, W. F., Ketchson, G. L. Predicting downslope travel of granitic sediments from forest roads in Idaho. Water Resources Bulletin 1996; 32:371-382.

Megahan, W. F., Kidd, W. J. Effects of logging and logging roads on erosion and sediment deposition from steep terrain. Journal of Forestry 1972; 7:136-141.

Miller, S. N., Guertin, D. P., Syed, K. H., Goodrich, D. C. "Using High Resolution Synthetic Aperture Radar for Terrain Mapping: Influence on Hydrologic and Geomorphic Investigations." In *Wildland Hydrology*, D. S., Olsen, Potyondy, J. P., eds. Herndon, VA: American Water Resources Association, TPS-99-3, 1999. http://www.esri.com.

Montgomery, D. R., Dietrich, W. E. A physically based model for the topographic control on shallow landsliding. Water Resources Research 1994; 30:1153-1171.

Nazarea, V., Rhoades, R., Bontoyan, E., Flora, G. Defining indicators which make sense to local people: Intra-cultural variation in perceptions of natural resources. Human Organization 1998; 57:159-170.

Neary, D. G., "Changing Perceptions of Watershed Management from a Retrospective Viewpoint." In *Land Stewardship in the 21st Century: The Contributions of Watershed Management*, P. F. Ffolliott, M. B. Baker, Jr., C. B. Edminster, M. C. Dillon, and K. L. Mora, tech. coords. Fort Collins, CO: Rocky Mountain Research Station, USDA Forest Service, Proceedings RMRS-P-13, 2000, pp. 167-176.

Norris, L. A., Lorz, H. W., Gregory, S. V. "Forest Chemicals." In *Influences of Forest and Rangeland Management on Salmonid Fishes and Habitats*, W. R. Megahan, ed. Bethesda, MD: American Fisheries Society, Special Publication 19, 1991, pp. 207-296.

Ogrosky, H. O., Mockus, V. "Section 21: Hydrology of Agricultural Lands." In *Handbook of Applied Hydrology: A Compendium of Water-Resources Technology*, V. T. Chow, ed. New York: McGraw-Hill Book Company, 1964, pp. 21-1-21-97.

Packer, P. E. Criteria for designing and location of logging roads to control sediment. Forest Science 1967; 13:1-18.

Pavari, A., *Forest Influences*. Rome: Food and Agriculture Organization of the United Nations, Forestry and Forest Products Studies 15, 1962.

Postel, S., *Last Oasis: Facing Water Scarcity*. New York: W. W. Norton Company, 1992.

Quinn, P., Beven, K., Chevallier, P., Planchon, O. "The Prediction of Hillslope Flow Paths for Distributed Hydrologic Modeling Using Digital Terrain Models." In *Terrain Analysis and Distributed Modelling in Hydrology*, K. J. Beven and I. D. Moore, eds. New York: John Wiley & Sons, Inc., 1991, pp. 63-84.

Quinn, R. M., Brooks, K. N., Ffolliott, P. F., Gregersen, H. M., Lundgren, A. L., *Reducing Resource Degradation: Designing Policy for Effective Watershed Management*. Washington, DC: U.S. Department of State, EPAT Working Paper 22, 1995.

Reid, L. M., *Research and Cumulative Watershed Management Effects*. Berkeley, CA: Pacific Southwest Forest and Range Experiment Station, USDA Forest Service, General Technical Report PSW-GTR-141, 1993.

Reid, J., Whittlesey, S., *The Archeology of Ancient Arizona*. Tucson, AZ: University of Arizona Press, 1997.

Rich, L. R., *Erosion and Sedimentation Movement Following a Wildland fire in a Ponderosa Pine Forest of Central Arizona*. Fort Collins, CO: Rocky Mountain Forest and Range Experiment Station, USDA Forest Service, Research Note 76, 1962.

Rich, L. R., Thompson, J. R., *Watershed Management in Arizona's Mixed Conifer Forests: The Status of Our Knowledge*. Fort Collins, CO: Rocky Mountain Forest and Range Experiment Station, USDA Forest Service, Research Paper RM-130, 1974.

Rickson, R., Saffigna, P., Vanclay, F., McTanish G. "Social Basis for Farmers' Responses to Land Degradation." In *Land Degradation: Problems and Policies*, A. Chisholm and R. Dumsday, eds. Sydney, Australia: Cambridge University Press, 1995, pp. 187-199.

Roller, N., Bergen, K. Integrating data and information for effective forest management. Journal of Forestry 2000; 98(6):61-63.

Sample, V. A., ed., *Remote Sensing and GIS in Ecosystem Management*. Washington, DC: Island Press, 1994.

Satterlund, D. R., Adams, P., *Wildland Watershed Management*. New York: John Wiley & Sons, Inc., 1992.

Savabi, M. R., Flanagan, D. C., Hebel, B., Engel, B. A. Application of WEPP and GIS-GRASS to a small watershed in Indiana. Journal of Soil and Water Conservation 1995; 50:477-484.

Schott, J. R., *Remote Sensing: The Image Chain Approach*. New York: Oxford University Press, 1997.

Sheng, T. C., Barrett, R. E., Mitchell, T. R. Using geographic information systems for watershed classification and rating in developing countries. Journal of Soil and Water Conservation 1997; 52:84-89.

Shortle, W. C., Smith, K. T. Aluminum-induced calcium deficiency syndrome in declining red spruce. Science 1988; 240:1017-1018.

Sidle, R. C., "Watershed Challenges for the 21st Century: A Global Perspective for Mountainous Terrain." In *Land Stewardship in the 21st Century: The Contributions of Watershed Management*, P. F. Ffolliott, M. B. Baker, Jr., C. B. Edminster, M. C. Dillon, and K. L. Mora, tech. coords. Fort Collins, CO: Rocky Mountain Research Station, USDA Forest Service, Proceedings RMRS-P-13, 2000, pp. 45-56.

Sidle, R., Pearce, A. J., O'Loughlin, C. L. Hillslope Stability and Land Use. American Geophysical Union, Washington, DC: Water Resources Monograph 11, 1985.

Steen, H. K., *The U.S. Forest Service: A History*. Seattle, WA: University of Washington Press, 1976.

Sturges, D. L., *Hydrologic Relationships on Undisturbed and Converted Big Sagebrush Lands: The Status of Our Knowledge*. Fort Collins, CO: Rocky Mountain Forest and Range Experiment Station, USDA Forest Service, Research Paper RM-140, 1975.

Swank, W. T., Tilley, D. R. "Watershed Management Contributions to Land Stewardship: Case Studies in the Southeast." In *Land Stewardship in the 21st Century: The Contributions of Watershed Management*, P. F. Ffolliott, M. B. Baker, Jr., C. B. Edminster, M. C. Dillon, and K. L. Mora, tech. coords. Fort Collins, CO: Rocky Mountain Research Station, USDA Forest Service, Proceedings RMRS-P-13, 2000, pp. 93-108.

Swift, L. W., Jr. Filter strip widths for forest roads in the southern Appalachians. Western Journal of Applied Forestry 1986; 10:27-34.

Tapscott, D., *Creating Value in the Network Economy*. Cambridge, MA: Harvard Business School Press, 1999.

The World Bank, *Water Resources Management*. Washington, DC: The World Bank, 1993.

Tiscareno-Lopez, M., Weltz, M. A., Lopes, V. L. Assessing uncertainies in WEPP's soil erosion predictions on rangelands. Journal of Soil and Water Conservation 1995: 50:512-516.

Tomlin, C. D., *Geographic Information Systems and Cartographic Modeling*. Englewood Cliffs, NJ: Prentice-Hall, 1990.

Toupal, R. S., Johnson, M., *Conservation Partnerships: Indicators of Success*. Tucson, AZ: Social Sciences Institute, USDA Natural Resources Conservation Service, Technical Report 7.1, 1998.

Troendle, C. A. The potential for water augmentation from forest management in the Rocky Mountain region. Water Resources Bulletin 1983; 19:359-373.

Troendle, C. A., King, R. M. The effect of timber harvesting on the Fool Creek Watershed, 30 years later. Water Resources Research 1985; 21:1915-1922.

Troendle, C. A., Wilcox, M. S., Bevenger, G. S. The Coon Creek water yield augmentation pilot project. Proceedings of the 66th Western Snow Conference; April 20-23; Snowbird, Utah. Fort Collins, Colorado, 1988, pp. 123-130.

Twigg, D. R. "The Global Positioning System and Its Use for Terrain Mapping and monitoring." In *Landform Monitoring, Modeling, and Analysis*, S. N. Lane, N., Richards, K., Chandler, J., eds. Chichester, England: John Wiley & Sons, Inc., 1998, pp. 37-61.

U.S. Environmental Protection Agency., *The Quality of Our Nation's Water*. Washington, DC; U.S. Environmental Protection Agency, EPA 841-S-94-002, 1992.

Van Haveron, B. P. A reevaluation of the Wagon Wheel Gap watershed experiment. Forest Science 1988; 34:298-214.

Verry, E. S., Hornbeck, J. W., Todd, A. H. "Watershed Research and Management in the Lake States and Northeastern United States." In *Land Stewardship in the 21st Century: The Contributions of Watershed Management*, P. F. Ffolliott, M. B. Baker, Jr., C. B. Edminster, M. C. Dillon, and K. L. Mora, tech. coords. Fort Collins, CO: Rocky Mountain Research Station, USDA Forest Service, Proceedings RMRS-P-13, 2000, pp. 81-92.

Warren, S. D. An erosion-based land classification system for military installations. Environmental Management 1989; 13:251-257.

Whitehead, P. G., Robinson, M. Experimental watershed studies - An international and historical perspective of forest impacts. Journal of Hydrology 1993; 145:217-230.

Wolinsky, A., *The History of the Internet and World Wide Web*. Springfield, NJ: Enslow Publishing Company, 1999.

Wright, K. A., Sendek, K. H., Rice, R. M., Thomas, R. B. Logging effects on streamflow: Storm runoff at Caspar Creek in northwestern California. Water Resources Research 1990; 26:1657-1667.

Yakowitz, D. S., Szidarovszky, F. Multi-attribute decision making: Dominance with respect to an importance order of the attributes. Applied Mathematics and Computation Computations 1993; 54:167-181.

Yakowitz, D. S., Lane, L. J., Szidarovszky, F. Multi-Attribute Decision Making: Dominance with Respect to an Importance Order of the Attributes. Applied Mathematics and Computation 1993; 54:167-181.

Yakowitz, D. S., Lane, L. J., Stone, J. J., Heilman, P., Reddy, R. K. "A Decision Support System for Water Quality Modeling." In *Water Resources Sessions/Water Form '92*, Mohammad Karamouz, ed. Baltimore, MD: American Society of Civil Engineering, Environmental Engineering Division, 1992a, pp. 188-193.

Yakowitz, D. S., Lane, L. J., Stone, J. J., Heilman, P., Reddy, R. K., Iman, B. "Evaluating Land Management Effects on Water Quality Using Multi-Objective Analysis with a Decision Support System." In *Ground Water Ecology: Proceedings of the 1st International Conference*, Jack A. Stanford and John J. Simons, ed. Bethesda, MD: American Water Resources Association, 1992b, pp. 365-374.

Yoder, D., Lown, J. The future of RUSLE: Inside the new Revised Universal Soil Loss Equation. Journal of Soil and Water Conservation 1995; 50:484-489.

Young, R. A., Onstad, C. A., Bosch, D. D., Anderson, W. P. AGNPS: A nonpoint source pollution model for evaluating watersheds. Journal of Soil and Water Conservation 1989; 44:168-172.

4

ISSUES TO BE CONFRONTED

Society must confront many issues when implementing effective watershed management practices, projects, and programs to attain better land stewardship in the future. Global issues and issues in the United States are considered in this chapter to place the perspectives, problems, and programs presented in the preceding chapter into a context to appreciate the contributions of watershed management to land stewardship. Many of these issues concern the status and future of the watershed management planning process. This process is complex because of the physical, biological, and social interactions that are the foundation of watershed management (Brooks et al., 1992, 1994, 1997; Thorud et al., 2000). However, people have a responsibility to act together in society to conserve natural resources and, when necessary, preserve their integrity for future generations. The outcome of this participatory effort is the sustainable use of natural resources through watershed management.

4.1. SUSTAINABILITY

Sustainability can be defined either broadly or narrowly, although the definition used should specify the spatial and temporal scales being considered (Conway, 1985; Brown et al., 1987; Gregersen and Lundgren, 1990; Gregersen et al., 1998). Sustainability is the long-term production of economically viable and socially acceptable goods and services, with a minimum of natural resource depletion and environmental deterioration. Sustainable watershed management programs provide goods and services for the short-term welfare of humans, while protecting the environment and maintaining the natural-resource base for use by future generations (Eckman et al., 2000). Sustainability is integrated with conservation through a watershed management approach to land stewardship. The global issues and issues in the United States considered below are considered within the concept of sustainability.

4.2. GLOBAL ISSUES

Scarcities of land, water, and other natural resources—and human responses to these scarcities—threaten the conservation, sustainable development, and use of these resources. These scarcities represent paramount environmental concerns worldwide for the coming century (Scherr and Yadav, 1996; Rosegrant, 1997). Another major concern is developing ways to cope with the extremes and uncertainties of global, national, and regional weather patterns and subsequent climatic conditions. The effectiveness of future land stewardship will depend on the severity of these extremes and the magnitudes of these uncertainties. Watershed managers must be aware of the possible impacts of changing weather patterns and climatic conditions on the sustainability of their planned practices, projects, and programs. It is within this context that Brooks and Eckman (2000) and Sidle (2000) suggest that watershed management provides a pragmatic perspective and framework to apply the technologies needed to help solve problems such as:

- Increasing water scarcity;
- Mitigating water pollution problems;
- Minimizing the effects of droughts, floods, and torrents;
- Increasing occurrences of landslides;
- Attaining a better understanding of the hydrologic response of headwater areas;
- Cumulative watershed effects;
- Scarcity of land and other natural resources; and
- Land and resource tenure.

Although these global issues are not all inclusive nor prioritized by importance or severity, the participants of the conference on "Land Stewardship in the 21st Century: The Contributions of Watershed Management," believe that they show the diversity of issues that future planners, managers, and decision-makers will confront (Ffolliott et al., 2000).

4.2.1. Increasing Water Scarcity

Water scarcity has been called the top global issue in the coming century by Engelman and LeRoy (1993), The World Bank (1993), Rosegrant (1997), Brooks and Eckman (2000), Gregersen et al. (2000), and others. About 50 countries, with an aggregate population of about 3 billion people, will suffer from some form of water scarcity in the next 25 years (Postel, 1992; Gardner-Outlaw and Engelman, 1997). A combination of factors causes water scarcity (The World Bank, 1993; Thomas et al., 1993; U.S. Agency for International Development, 1993) including:

- Growing demands for limited water supplies;
- Inefficient use of existing water supplies;
- Unequal water distribution; and
- Degradation of water quality.

4.2.1.1. Scarcity Resulting from Limited Water Supplies

Demands for limited water supplies intensify the number of people and their livestock increase and as more people move from rural into urban areas where water demands are concentrated (Box 4.1). While technological breakthroughs, such as installation of desalinization plants, can increase the usable water resources in affluent countries (U.S. Agency for International Development, 1993), they are costly and small in scale.

Box 4.1 Water Scarcity in the Middle East: A Crisis Situation

The scarcity of water in the Middle East could soon reach crisis proportions. People living in the region currently consume more fresh water than is produced naturally within their boundaries (Metz, 2000). A few examples illustrate this crisis.

The residents of Amman, Jordan, a city of more than 1 million people, are allowed to have running water only once a week. The Quwayq River, once the main water source for Syria's second largest city, Aleppo, has been pumped dry. Seventy percent of Iraq's agricultural crops perished in 1999. In the southernmost city of Basra, a system of canals from the Tigris and Euphrates Rivers has been reduced to polluted trickles. In 1999, the Sea of Galilee, a freshwater lake that supplies one-third of Israel's water, dropped below the demarcation point—the level that water quality becomes threatened intruding saline water.

These and other water-scarcity problems in the Middle East are acute and contributing to public health problems, while placing severe limits on regional economic growth. The seriousness of allocating the limited water supplies among competing sectors within the countries of the Middle East are overshadowed by the problems of allocating water across national boundaries. Few agreements exist about how to manage the regional water sources and conflicts could arise as users claim greater volumes of water.

4.2.1.2. Inefficient Use of Existing Water Supplies

One reason that water supplies are ineffective at meeting needs, is their careless use. Agriculture has frequently been targeted in this regard because it often uses water ineffectively (Rosegrant, 1997; The World Bank, 1993; U.S. Agency for International Development, 1993). Substantial water scarcities in urban areas occur through losses from leaks in distribution systems, on-site waste, and waste at uncontrolled community taps. Some water leakage and waste are lost through evaporation, some seeps into the ground, and some can be recovered downstream, but all occur at a significant cost in wasted pumping and treatment.

4.2.1.3. Inequitable Water Distribution

Innovative technological approaches alone are sometimes insufficient in solving water scarcity problems. The unequal distribution of water can be more limiting than a failure to implement new or improved technologies (U.S. Agency for International Development, 1993). The lack of institutional capabilities to plan and manage for scarce water supplies and a failure to incorporate market forces into water allocation plans can also contribute to this problem.

4.2.1.4. Degradation of Water Quality

In many countries, people are confronted with the continuing degradation of water quality or increasing water pollution (see below). Available water-use options usually narrow as the degradation of water increases, placing many people at greater future risk (The World Bank, 1993; U.S. Agency for International Development, 1993). The self interests of local water users can compound the problem of water-quality degradation.

These people focus on short-term benefits, while sacrificing the long-term integrity of the often limited water resource.

4.2.2. Water Pollution Problems

Point and nonpoint water pollution problems continue to plague many countries. Increasing water pollution jeopardizes the sustainability of goods and services derived from watershed lands, threatens the health of people and their livestock, compounds water scarcity, adversely effects the functioning of aquatic ecosystems, and ultimately impacts the economy of the region and country (Engelman and LeRoy, 1993; U.S. Agency for International Development, 1993; The World Bank, 1993; Thomas et al., 1993; Brooks and Eckman, 2000).

Major sources of surface and groundwater degradation are:

- Increasing discharges of untreated or inadequately treated wastewater;
- Emissions from agro-processing plants and unregulated agrochemical use; and
- Discharge of hazardous, toxic, or other industrial waste into water supplies.

Other factors contributing to a decline in water quality in many countries are drainage of saline water from agricultural lands and an overdraft of groundwater resources that result in saline intrusions from the sea. The degradation of water quality is often blamed on local-level mismanagement, but its origins are more often the result of improper policy decisions addressing economic and soil problems at the regional, national, or international level.

4.2.2.1. Irrigation Water

Increased use of chemicals in many irrigation practices can result in pollutant emission into water resources. Therefore, forest, grassland, or agricultural cropland irrigation can become a serious pollution problem by increasing salt concentrations and adding toxic organic materials to the drainage water from irrigated lands. A decline in water quality is especially problematic when drainage water containing nitrates, other soluble salts, and when pesticides percolate through the soil (Yaron et al., 1973; Ayers and Westcott, 1976; Armitage, 1987; U.S. Agency for International Development, 1993). Sediments in surface water from irrigation can also contain minerals and pesticides that adhere to the soil particles. These sediments are often deposited in slow moving streams or still water bodies where they can adversely affect wildlife and aquatic organisms. Polluted irrigation drainage water can also severely impair downstream irrigation systems.

4.2.2.2. Appropriate Technologies

A major issue confronting countries impacted by point and nonpoint pollution is how to take advantage of the opportunities for applying appropriate "clean" technologies to prevent further water-supply contamination (Engelman and LeRoy, 1993; U.S. Agency for International Development, 1993). Using low input and sustainable techniques in natural resource and agricultural management can reduce reliance on pesticides and fertilizers, while maintaining the production level and lower management costs. To safeguard water quality, urban industries can implement the following:

- Reducing toxic inputs;
- Reducing non-product waste outputs by improving manufacturing practices; and
- Manufacturing and using environmentally responsible products.

4.2.3. Droughts, Floods, Torrents

Droughts, floods, and torrents are natural phenomena that cause devastating damage to people and infrastructures. Effected countries and responsible agencies often make large investments from limited budgets to cover the costs of mitigating the effects of these events (The World Bank, 1993; Keeney and Alexander, 1994).

4.2.3.1. Droughts

Droughts are departures from average conditions in which a shortage of water adversely affects ecosystem functioning and the well being of people and their livestock. Characterized by shortages of water and food for people and forage for livestock, droughts often lead to unplanned use of natural and agricultural resources (The World Bank, 1985; 1993; Ffolliott et al., 1995; Chandra and Bhatia, 2000). If contingency planning is not undertaken to meet these shortages, large-scale degradation of land, water, and other natural resources results. While droughts will continue to occur in the dryland regions of the world, predicting when they will occur or their duration is impossible. Human population growth and possible climatic changes compound the problems of desertification[19] and threaten the integrity of dryland ecosystems and the success of traditional land-use practices in drought-prone regions.

4.2.3.2. Floods and Torrents

Annually, floods and torrents cost billions of dollars worldwide for prevention and forecasting (Brooks and Eckman, 2000). However, it is the loss of lives and property damage due to these naturally occurring phenomena that are staggering. People's encroachment on flood plains and other hazardous areas exacerbates the impacts of these events. This encroachment is often due to land scarcity. In many countries, people have relied on dams, levees, and channel structures in river basins, along flood plains, and in areas susceptible to debris torrents. These structural solutions impart a false sense of security to those living in the hazardous areas. Leopold emphasized this point (1994) in his assessment of the devastating flooding along the Mississippi River in 1993.

Over many decades, much of the Mississippi River and its tributaries have been altered with locks and dams, levees, and channel and flood-plain modifications. However, the magnitude of the 1993 flood revealed the limitation of people's ability to engineer river systems to prevent flood damage. Replacement of natural wetlands, riparian areas, and flood plains with agricultural and urban systems over the years can cumulatively add to downstream flooding problems. Maintaining a watershed perspective brings these possible cumulative effects into focus.

---

[19] The process of becoming arid land or desert usually from land mismanagement or climate change.

4.2.4. Landslide Occurrences

Timber harvesting, extensive vegetative conversions, and improper road construction has increased the occurrence of landslides that result in life and property losses in many mountainous countries (Swanson and Dryness, 1975; Megahan et al., 1978; Brooks et al., 1997; Sidle, 2000). Processes that influence this increase in landslide activity vary with the type of disturbance. Increases in the occurrence of landslides have been observed 3-to-15 years after timber harvesting in many regions worldwide (Bishop and Stevens, 1964; Swanson and Dryness, 1975; O'Loughlin and Pearce, 1976). Landslides occur after logging when significantly reduced root strength exists and after a major storm or rapid snowmelt. Conversion of forests, woodlands, and shrubland vegetation to grasslands also reduces rooting strength in the soil and, consequently, landslide frequency and volume increases (Rice et al., 1969; O'Loughlin and Pierce, 1976). Slash-and-burn agricultural practices—common in many Latin American and Asian countries—can reduce site stability when steep watershed lands that support forest covers are converted to temporary croplands, with weak root-strength characteristics.

4.2.4.1. Impacts of Roads

Roads and road networks that are improperly designed, poorly constructed, insufficiently maintained, or on steep terrain are the largest contributors of landslide erosion on a per-unit-area basis (O'Loughlin and Pierce 1975; Sidle et al., 1985; Megahan and Hornbeck, 2000). Stability problems associated with roads include overloading effects on the embankment fill material, placement of unstable fill materials on steep slopes, undercutting the hillslope, and redirecting road drainage onto unstable sections of the hillslope. Improper road drainage is commonly blamed for many road-related failures (Sidle, 2000). The magnitude of this problem is difficult to predict due to the complex nature of drainage systems, imperfect knowledge of road hydrology, and problems associated with clogged cross drains and other drainage-system failures during runoff events.

4.2.4.2. Predicting Landslide Occurrence

There has been only limited success in predicting where landslides will occur, what the downslope or downstream impacts will be, or how various watershed management activities will affect the probability of slope failures (Sidle, 2000). At the landscape level, however, terrain-evaluation procedures are available to relate broadly-based categories of landslide-hazard estimates to timber harvesting, road building, and other management activities (Howes and Kenk, 1988). In regions where high-quality site data and historical landslide records are available, the effects of land use on slope failures have been evaluated by weighted multi-factor overlays (Nielson et al., 1979). However, these procedures are mostly qualitative and their successful application relies on expert knowledge.

Potential exists for improving on these terrain-evaluation procedures. One possibility is to include the weighted factors that reflect terrain attributes associated with landslides and emulate the underlying processes that contribute to slope failure in the prediction procedures (Sidle, 2000). Causative forces, such as rainfall intensity and duration, seismicity, snowmelt, and parameters influencing landslide potential (root strength, slope gradient, topographic expression, and groundwater concentration zones), should be incorporated into terrain-evaluation procedures. Another needed improvement in these procedures is the application of stability-assessment methods to larger geographic areas or areas that experience slump earthflows, debris avalanches, or other types of multiple failure.

4.2.4.3. Relationships Between Processes

Another poorly understood topic related to terrain stability is the relationship among earthflows, debris avalanches, and other hillslope and main-channel processes—debris flows, transport of suspended sediment and bedload, and channel scour and fills (Brooks et al., 1997; Sidle, 2000). Understanding this relationship is essential to predict the long-term effects of vegetation management on aquatic habitats, fluvial geomorphology, and water quality. While low-gradient downstream reaches have been studied regarding sediment movement, hydrologic response, and aquatic productivity, headwater systems have largely been ignored. Headwaters situated on steep terrain are subject to shallow landslides, debris flows, bank failures, and other active erosion processes. Woody debris in headwater channels often provides temporary storage sites for this sediment, although the dynamics of the sediment storage and release in relation to the woody debris is poorly understood.

Management of riparian ecosystems in headwater systems has been intensely scrutinized. However, little long-term data supports the various economic, environmental, and political objectives. Therefore, issues, such as the width of buffer-leave strips necessary to protect channels and supply a sustainable level of large woody debris to the channels, continue to be debated (Streeby, 1971; Murphy and Koski, 1989; Sidle, 2000). Furthermore, the effects of changes of woody debris inputs over an entire forest rotational cycle (up to 100 years) on the overall hydrologic attributes of headwater systems are unknown, particularly with respect to sediment movement, channel condition, and aquatic habitats.

4.2.5. Hydrologic Response of Headwaters

Watershed managers should understand the hydrologic response of forested and other headwater systems in relation to the type of land-use practices that ensure the sustainability of the water flows from upland watersheds. Various forested headwater features can cause different hydrologic responses in comparison to the headwaters of agricultural or urban watersheds and river basins that have a mixture of intermingling land uses. For example, most forest soils have relatively high infiltration capacities and, consequently, overland water flow rarely occurs. This situation is especially true in naturally regenerated temperate, subtropical, and tropical forests where a substantial amount of soil organic matter is found. Subsurface water flow plays either an active role in stormflow generation at these headwater sites or a passive role in recharging riparian areas (DeBano and Schmidt, 1989; Sidle, 2000; Verry et al., 2000a). Confounding these roles is that fact that headwater sites are susceptible to compaction and other disturbances from ongoing land-use activities. Certain types of plantation forests can promote overland flow due to the exclusion of herbaceous plant species in the understory and the lack of organic litter.

4.2.5.1. Hydrologic Functioning

Issues to address in understanding the hydrologic response in headwater systems include specifying the flow mechanisms or pathway functioning at different spatial scales within a watershed system. Original research behind the variable-source-area concept of streamflow generation from watershed lands (Hewlett and Hibbert, 1967; Dunne and Leopold, 1978; Satterlund and Adams, 1992; Brooks et al., 1997) was conducted in the mountains of the southeastern United States. Research should also be conducted on sites with gentle slopes, differing subsurface-flow regimes, and changing groundwater hydrology. Insights into hydrogeomorphic relationships are also needed to clarify the spatial and temporal attributes of flow paths that affect headwater and downstream systems, including the impacts of land use (Sidle and Hornbeck, 1991; Burgess et al., 1998). Understanding the dynamics of flow paths in headwater systems related to changing antecedent moisture conditions, topographic attributes, and watershed-management impacts are important (Sidle, 2000).

4.2.5.2. Cause-and-effect Relationships

Cause-and-effect relationships between hydrologic functioning and geomorphic attributes require further investigation. Factors influencing threshold responses of hydrologic functions, such as runoff from hillslope hollows, expansion of preferential flow networks, and redistribution of subsurface water storage, also needs additional study. These thresholds could have different scale dependencies within the range of zero to second-order watersheds (Sidle, 2000). Another issue of concern related to hydrologic response is the routing of water from headwater to lower-gradient channels. Woody debris, boulders, and other roughness elements have a greater influence on water routing in headwater channels than on larger downstream systems (Abbe and Montomery, 1996). Watershed management practices related to the dynamics of woody debris and hillslope processes often influence hydrologic routing and cumulative watershed effects.

4.2.6. Cumulative Watershed Effects

A variety of land uses are typically distributed on larger watersheds according to natural resource availability, site-productivity level, ownership or tenure, and land-use restrictions (Sidle, 2000). The spatial distribution of the land uses often changes through time depending on economic conditions, environmental issues, ownership, technology, and regulatory constraints. These spatially and temporally distributed effects can interact with each other and the natural ecosystem processes to produce cumulative effects on watershed resources (Sidle and Hornbeck, 1991; Reid, 1993; Brooks et al., 1997). Global climate change and changing local and regional demographics also contribute to cumulative watershed effects.

4.2.6.1. On- and Off-site Effects

Effected natural resources can be on- or off-site of the actual land-use impacts (Reid, 1993; Brooks et al., 1997; MacDonald, 2000; Sidle, 2000). On-site effects include those on vegetation, soil, and nutrient cycling. Changes from cumulative on-site effects influence the long-term productivity of a watershed. The changes can also have major impacts on water

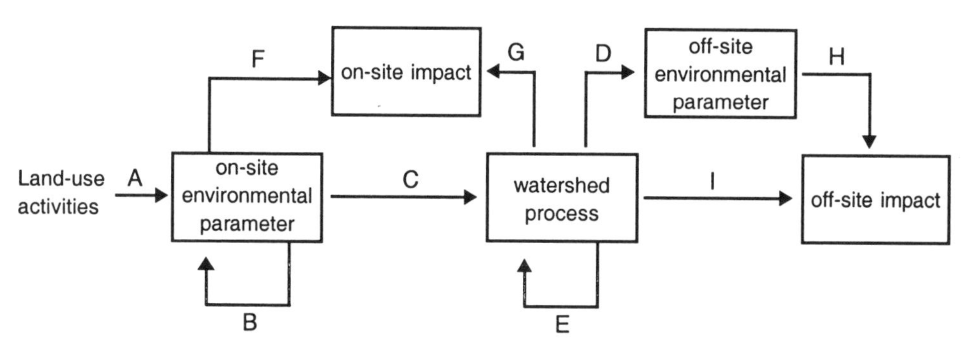

Figure 4.1. Influences that generate on-site and off-site cumulative watershed effects (adapted from Reid, 1993). The interaction paths are described in the text.

production, timber management, rangeland use, and recreational opportunities. Each cumulative off-site watershed effect occurs because the environmental change alters the production or transport of water, sediment, organic matter, chemicals, and heat. When the responses from many sites are combined and transported to a common site or when a transported response interacts with an on-site change at another site, off-site effects occur.

Considering the mechanisms that trigger impacts helps to understand the nature of on- and off-site effects (Reid, 1993). Land-use activities can influence only a few environmental parameters (Figure 4.1, path A). Changes in vegetation, soil characteristics, water quality, and fauna can induce compensatory changes (Figure 4.1, path B) and influence watershed processes (Figure 4.1, path C). Watershed processes arise from a watershed's role as a concentrator of runoff and include production and transport of runoff, sediment, chemicals, organic materials, and heat. These processes can influence environmental parameters (Figure 4.1, path D) and interact with each other (Figure 4.1, path E). Changes in either watershed processes or environmental parameters can generate on-site effects (Figure 4.1, path F and G), while only changes in watershed process can produce off-site effects (Figure 4.1, path H and I).

The properties of the impacted watersheds and ecosystems have complicated cumulative watershed effects that obscure the cause-and-effect relationships (Reid, 1993; Brooks et al., 1997). The effects can be delayed until long after the triggering impacts have occurred and the location of these impacts is frequently far-removed from the original land-use disturbance (Box 4.2). Local conditions can also modify the form of an impact and a single impact can have many contributing causes.

Box 4.2 Cumulative Watershed Effects: On- and Off-Site Examples

Examples of cumulative watershed effects that occur on-site include increased landslide susceptibility due to repeated timber harvesting (Sidle, 1991), progressively worse gully development in response to the timber harvesting (Prosser and Soufi, 1998), and increases in soil compaction and overland flow (Warren et al., 1986). Among the commonly encountered off-site effects are the alterations of channel morphology and sedimentation regimes (Lyons and Beschta, 1983; Sidle and Sharma, 1996), changes in water-quality constituents (Boyer and Parry, 1987; Sidle and Amacher, 1990), riparian-vegetation responses (Kauffman and Kreuger, 1984; Potts, 1998; Verry et al., 2000a), and stormflow changes (Jones and Grant, 1996; Thomas and Megahan, 1998).

4.2.6.2. Assessment of Effects

An assessment of possible cumulative watershed effects is critical to holistic watershed management and, therefore, a watershed management approach to land stewardship. This assessment requires an improved understanding of how water, sediments, nutrients and organic materials, and pollutants are routed through a complexity of landscapes and what changes occur along the way (Loftis and MacDonald, 2000; MacDonald, 2000; Sidle, 2000). Understanding these routing processes requires consideration of:

- Spatial and temporal scaling issues, such as the hydrologic thresholds, which initiate stormflow (Brooks et al., 1997);
- Process relationships (Sidle et al., 1995; Brown et al., 1999);
- Spatial variability of landscape properties (Sinowski and Auerswald, 1999); and
- Disaggregation and aggregation in hydrologic behavior (Becker and Braun, 1999).

The role of episodic events is also important in assessing cumulative watershed effects. These events define thresholds of concern for ecosystem processes (Reid, 1993; Sidle, 2000). Occurrences of disturbance events above or below these thresholds can determine the magnitude and duration of the cumulative watershed effects.

4.2.7. Scarcity of Land and Other Natural Resources

Worldwide scarcity of land and other natural resources results from a shrinking arable land base caused by expanding populations of humans and their livestock. Land degradation occurring from improper agricultural cultivation, excessive livestock grazing, and extensive deforestation of marginally productive lands compounds the effects of land scarcity (Engelman and LeRoy, 1995; Gardner-Outlaw and Engelman, 1999; Brooks and Eckman, 2000). Effected lands are often steep sites that experience accelerated surface and gully erosion, soil mass movement, and increased stormflow volumes and sediment damage.

It is estimated that about 0.5 ha of farmland is needed to support one person living in the tropics (Pimental et al., 1995). It is also estimated that by the year 2025, about 50 tropical countries will have less than 0.1 ha of arable land per capita (Lal, 1997). Of the total land base of forests, woodlands, and rangelands found globally, between 20% and 25% have been degraded since the 1950s, with nearly 4% of these lands being severely degraded (Scherr and Yadav, 1996).

Deforestation and desertification of productive watershed lands are two issues that continue to gain worldwide attention (Repetto, 1985; Schlesinger et al., 1990; Apelt et al., 1993; Anaya-Garduno et al., 1994; Maser, 1994; Mouat and Hutchinson, 1995; Squires and Sidahmed, 1998). Consequences of deforestation and desertification include:

- Soil degradation and losses in land productivity;
- Loss of biological diversity; and
- Inability of vegetative cover to protect lands.

Equally important are the effects of deforestation and desertification on the watershed functions that supply high-quality water to people and their livestock in a sustainable and environmentally-sound way.

4.2.8. Tenure Issues

Land and resource tenure and the associated rights of tenure are often institutional constraints to sustainable watershed management and effective land stewardship. Frequently, land and resource tenure are neglected when planning watershed practices, projects, and programs (Gregersen et al., 1987; Ffolliott, 1995; Brooks et al., 1997; Brooks and Eckman, 2000). To attain the desired level of sustainability in these efforts and, ultimately, sustainable development, this planning process must acknowledge the tenure arrangements on the watershed lands involved.

Appropriate studies of local institutional arrangements should be conducted early in the planning process. For example, when a central government owns trees and forests, planners of watershed practices, projects, and programs that promote reforestation should understand and appreciate how these activities could affect the people living on the watersheds and how these people might respond to the planned reforestation efforts. Questions such as, "Who has the rights to the planted trees and what are the methods available to solve conflicts over rights?" must be resolved to the satisfaction of all the parties. When land and resource tenure arrangements are barriers to planned watershed management goals, institutional or policy changes are necessary.

4.3. ISSUES IN THE UNITED STATES

One of the most central issues of watershed management in the United States will continue to be the demand for high-quality water (Brooks and Eckman, 2000; de Steiguer, 2000; Thorud et al., 2000). The availability, characteristics and properties, and behavior of water in natural ecosystems depends on the watershed from which the water is derived and the prevailing climate and land-use practices on the watershed lands involved. Watershed management will become increasingly significant as a way to sustain the supply of high-quality water for a variety of uses. High-quality water that is readily available in adequate quantity for human use will become an increasingly prominent consideration in watershed management—it will likely be viewed as a human health and scarcity issue.

Other watershed-management issues important to people in the United States include:

- Maintaining water-quality standards;
- Appreciating the role of prescribed fire in watershed management;
- Mitigating the impacts of roads and trails on hydrologic processes;
- Understanding the implications of watershed relationship to management;
- Recognizing the impacts of urbanization on the hydrologic processes; and
- Developing best management practices (BMPs) for a wide range of watershed conditions.

Although these issues are not necessarily all inclusive nor prioritized by importance or severity, the participants at the conference on "Land Stewardship in the 21st Century, The Contributions of Watershed Management" identified them as a high priority (Ffolliott et al., 2000).

4.3.1. Maintenance of Water-Quality Standards

Concerns about point and nonpoint water pollution have heightened in the United States due to the need to comply with water-quality standards specified by the federal and state government. Water-quality standards are established as part of determining the designated beneficial use of a water body (Lamb, 1985; Foster et al., 1995; Stumm and Morgan, 1995; Brooks et al., 1997). A water body can have a diverse array of designated beneficial uses such as water supply and environmental and recreational uses. Designation of beneficial uses into one of these categories is developed through an open, public comment process (Brown et al., 1993; Tracy et al., 2000), which is thought to be the most equitable approach available. Once the designated beneficial uses have been determined, the physical, chemical, and biological characteristics of the water in relation to the uses—the established water-quality standards—must be maintained.

4.3.1.1. Regulatory Considerations

The process of determining the designated beneficial uses of a water body can create an adversarial relationship between private landowners within the watershed and the public that might use the watershed as recreational resource. This is especially true on many Western watershed lands where many people recreate near or in the water bodies (Brown et al., 1993; Tracy et al., 2000). The land base within such a watershed might be owned by only a fraction of the population that uses the watersheds as a recreational resource. Maintenance of the designated water-quality standards could restrict the land development or the land-use practices of the private landowners. Preventing private landowners from developing their property or limiting their property use can be a legal issue. Highly restrictive land-development or land-use laws can cause extended litigation initiated by private landowners against the planning or regulatory agency that developed and enforces the water-quality standards.

4.3.1.2. Total Maximum Daily Loads

The problem of maintaining water quality is compounded by the need to develop total maximum daily load (TMDL) plans for a water body, with water-quality conditions that limit its ability to meet its designated beneficial uses as mandated by the Clean Water Act (Gelt, 2000; Tracy et al., 2000). TMDLs for a water body are the sum of the waste-load allocations from point and nonpoint pollution sources within the watershed. A TMDL plan for a listed water body should identify current waste loads to the water body and the water-load capacities of the water body.

With this information, the waste-load capacity is then allocated to the known point and non-point pollution sources within the water body's watershed. However, preparation of TMDL plans is difficult because understanding and analyzing the mode of conveyance of nonpoint sources is complex (Brooks et al., 1997). More conventional hydraulic methods can generally be used for monitoring and analyzing the discharge of pollutants from point sources, such as through a pipe, but both nonpoint and point pollution sources must be included in a complete TMDL plan.

4.3.2. Role of Prescribed Fire[20]

Ecologists and land managers are evaluating the feasibility of reintroducing fire into many natural ecosystems in the United States to:

- Reduce excessive fuel buildups to mitigate the occurrence of catastrophic wildland fires and
- Meet other land management objectives such as preparing receptive seedbeds, thinning overstocked timber stands, increasing forage production, improving wildlife habitats, and enhancing aesthetic environments.

However, the impacts from either a prescribed fire or a prescribed natural fire[21] are not always sufficiently known to successfully reintroduce fire into natural ecosystems with confidence (Tiedemann et al., 1979; Baker, 1990; Pyne et al., 1996; DeBano et al., 1998). The severity of the fire determines the site-specific impacts, therefore, they are often difficult to extrapolate from one situation to another. Nevertheless, making a few qualitative generalizations about the potential effects of prescribed fire on streamflow regimes, erosion and sedimentation, and water quality is possible.

4.3.2.1. Streamflow Regimes

Streamflow response to a prescribed fire is generally smaller in magnitude—and almost nonexistent in instances of a low-severity prescribed burn—in contrast to responses to catastrophic wildland fire. It is usually not the purpose of a prescribed fire to completely consume litter accumulations and other organic matter on soil surfaces, which would result in large alterations in streamflow regimes (Gottfried and DeBano, 1990; Robichaud and Waldrop, 1994; Zwolinski, 2000). The occurrence of fire can increase peak streamflows when it causes the destruction of vegetation and reduces litter and other decomposed organic matter (Anderson et al., 1976; Scott, 1993). Streamflow regimes from burned watersheds often respond to rainfall inputs faster than watersheds supporting a protective vegetative cover, producing events where the time-to-peak occurs earlier (Brooks et al., 1997).

4.3.2.2. Erosion, Sedimentation, and Water Quality

The layer of litter and other organic matter that remains after a prescribed burn protects sites from increased surface erosion (Baker, 1990; DeBano et al., 1997; Zwolinski, 2000).

[20] A management ignited wildland fire that burns under specified conditions where the fire is confined to a predetermined area and produces the fire behavior and characteristics to attain planned fire treatment and resource management objectives.

[21] A naturally ignited wildland fire that burns under specified conditions where the fire is confined to a predetermined area and produces the fire behavior and characteristics to attain planned fire treatment and resource management objectives.

An increase in soil exposure and, therefore, surface erosion following a fire, causes much of the eroded soil to move only a short distance downslope. Sedimentation following a prescribed burn is generally less than that following a catastrophic wildland fire (Copley et al., 1944; Baker, 1990; Brooks et al., 1997; DeBano et al., 1998). However, when sediment yields increase because of any type of burning, they are often highest in the first year after the fire and decline in subsequent years as protective vegetation becomes established on the watershed.

Water-quality responses to fire are most evident in sediment and turbidity[22]. Less is known about the effect of fire on turbidity than on sedimentation because measuring turbidity is difficult due to its highly transient and variable character (Wright et al., 1976; Tiedemann et al., 1979). Nevertheless, turbidity levels can increase following a fire because of ash and slit to clay-sized particles in the water. After a fire, the reduced vegetative cover also increases the susceptibility of a watershed to nutrient loss through erosion (Grier, 1975; DeBano et al., 1998). Soil-cycling mechanisms reduce the nutrient uptake opportunities of plants when a reduced vegetative cover exists. This action further increases the potential nutrient loss that occurs through leaching. Overland water flow that percolates through the litter layer to the soil surface contains fewer bacteria than water that had not passed through the litter layer (Brooks et al., 1997). Therefore, the destruction of the litter layer might result in higher concentrations of bacterial and other biological organisms flowing overland to a stream channel.

4.3.2.3. Gaining Public Support

To successfully meet land stewardship objectives requires careful planning and public support concerning the use of prescribed fire (Cortner et al., 1984; Taylor et al., 1986; Czech and Ffolliott, 1996). Watershed managers and fire specialists must recognize and effectively articulate any risks associated with prescribed burning to decision-makers and the public. One obstacle that must be overcome to gain public acceptance of prescribed fire use in the United States concerns the "Smokey Bear" syndrome. The fire-prevention campaign built around Smokey Bear in the past 50 years was almost too effective in emphasizing the danger posed by and damaging effects of wildland fire. Smokey Bear represents fire suppression, containment, and extinguishment. Today, fire-prevention campaigns in the United States are educating the public about fire's beneficial role in natural ecosystems. As a result, people are beginning to understand the important role that fire must be allowed to continue to play to maintain healthy ecosystems. Prescribed fire is a valuable natural resource management tool.

4.3.3. Impact of Roads and Trails

Roads may have undesired and negative effects on hydrology, geomorphic features, sedimentation, and water quality. Natural resource management can be little more than custodial without land access. Roads and trails are often necessary to implement watershed management (Box 4.3). However, the potential impact of these transportation corridors on erosion and the consequent sediment production often exceeds that caused by all other management activities combined (Satterlund and Adams, 1992; Brooks et al., 1997). No

---

[22] A measure of the optical property of water that scatters light.

Box 4.3 The Role of Roads in Natural Resource Management: An Example

The 7,200 km² Coronado National Forest in southeastern Arizona encompasses some of the most scenic and biologically diverse public lands in the United States. However, the public and responsible land managers are having difficulty accessing the forest (McGee, 2000).

When the West was being settled, transportation routes were critical. Most landowners welcomed new roads and often gave verbal permission for public use of roads located on their property. Because these roads were a benefit to all, little importance placed on recording the legal rights-of-way. Today, nearly two-thirds of all of the roads that access the Coronado National Forest have no legally documented right-of-way. The complexity of ownership compounds the access problem across the forest. State, private, and other federally managed lands surround the Coronado's 12 "sky island" mountain ranges. Forest neighbors are diverse blends of major urban areas, rapidly expanding cities and towns, commercial industrial lands, military reservations, individually-owned parcels of lands, and almost 100 kilometers of common boundary with the Republic of Mexico.

The increasing human population has pushed development of what were remote private lands to the boundary of the forest. Developers and owners of these private lands can legally block access to thousands of square kilometers of public lands by locking gates on their private roads. To help address the issues of access and road density, the Forest Service is currently inventorying all of the roads leading into and within the Coronado National Forest. This inventory will allow land managers to identify the legal and illegal roads and then decide how many and what kinds of roads should remain accessible or should be obliterated.

other activity results in such intensive and concentrated soil disturbance as does the construction and maintenance of roads and trails.

Improper road and trail design, layout, and construction cause many erosion and sediment problems (Megahan and Hornbeck, 2000), which can be eliminated in the planning stage. The issue, therefore, is not necessarily how to reduce erosion and sedimentation caused by road and trail systems. Instead, the concern is what should be done about the maintenance or improvement of the current system of roads and trails that traverse watershed land. These lands often encompass the headwaters of streams that converge and flow from upland watersheds.

4.3.3.1. Current Dilemma

National Forest System (NFS) roads illustrate the current road maintenance and improvement dilemma in the United States. Between 650,000 and 800,000 kilometers of roads exist on NFS lands (Bell, 2000). The Forest Service is responsible for 622,000 kilometers of these roads, with the remaining roads on easements approved by the Forest Service and maintained by states, counties, and private individuals (approximately 124,000 kilometers).

From 1991 through 1999, approximately 430,000 kilometers of roads were decommissioned[23] annually (USDA Forest Service, 1999). Road decommissioning involves

[23] Demolition, dismantling, removal, obliteration, or disposal of a deteriorated or otherwise unneeded asset or component, including necessary cleanup work. This action eliminates the deferred maintenance needs for the fixed asset. Portions of an asset or component may remain if they do not cause problems or require maintenance.

using various levels of treatments to restore unneeded roads to a more natural state to mitigate environmental damage and restore hydrologic function. An estimated 97,300 kilometers of unclassified[24] roads exist on NFS lands. Today, NFS roads serve a variety of forest users and join with county, state, and national highways to connect rural communities and urban center with NFS lands. Recreation is the single largest use or activity supported by NFS roads, accounting for approximately 90% of the daily traffic. Administrative use (9%) and commercial uses (1%) make up the balance (Coghlan and Sowa, 1998).

Public access to NFS lands by low-clearance passenger cars is limited. Although NFS roads are maintained to accommodate low-clearance passenger cars and high-clearance vehicles (sport-utility vehicles, pickups, and jeeps), only about 122,000 kilometers, or 20%, of NFS roads are maintained for low-clearance passenger cars. In contrast, 359,000 kilometers, or 57%, of NFS roads are designated and maintained for high-clearance vehicles. The remaining kilometers of NFS roads are single-use roads that are generally closed after their initial use and kept closed between uses (USDA Forest Service, 2000).

A major dilemma confronting the Forest Service is the growing backlog of deferred maintenance and capital improvement needs on NFS lands (Bell, 2000). A 1998 survey of road maintenance[25] and capital improvement needs on NFS lands showed an annual maintenance budget backlog of $8.4 billion. The deferred maintenance backlog alone was $5.5 billion or 66% of the total backlog. The total fiscal year 2000 road maintenance budget of $111 million—an $11 million increase over fiscal year 1999—will meet less than 20% of the agency's annual needs and less than 50% of identified critical needs. Each year's unmet maintenance increases the backlog as roads deteriorate and the cost of repairs continues to rise (USDA Forest Service, 2000).

Road maintenance also includes investments necessary for compliance with relevant state and federal statutes such as the Clean Water Act, National Environmental Policy Act, and the Endangered Species Act. Additionally, Forest Service policy requires that planning, designing, constructing, and maintaining NFS roads is accomplished using the best available scientific information and BMPs. Due to the growing backlog of deferred road maintenance on NFS lands, many roads are not in compliance with applicable laws and regulations.

4.3.3.2. Solutions to the Problems

One of the 4 elements in the Forest Service *Natural Resource Agenda* is roads (USDA Forest Service, 1998; Towns, 2000), and roads are integrally linked to the other elements in this agenda—clean water, recreation, and sustaining healthy ecosystems. For example:

- Roads negatively affect water quality;
- Recreation accounts for approximately 90% of the current use of NFS roads; and
- Management of natural ecosystems to sustain healthy vegetation is impossible without the access provided by roads.

[24] Roads on National Forest System lands that are not needed for, and not managed as part of, the forest transportation system, such as unplanned roads, abandoned travelways, and off-road vehicle tracks that have not been designated and managed as a trail, and those roads no longer under permit or authorization.

[25] The ongoing upkeep of a road necessary to retain or return the road to the approved road management objective.

The *Natural Resource Agenda* lists 4 actions that address the problem of reconciling the number of the kilometers of NFS roads and the funds available for their maintenance (Bell, 2000). These actions are:

- Implementing a roads policy to make existing NFS roads safe, environmentally sound, and affordable to manage;
- Accelerating the current pace of decommissioning roads;
- Selectively upgrading roads and improving or restoring key roads for recreation, rural access, and a sustainable flow of good and services; and
- Seeking additional funding sources for maintenance and capital improvement of NFS roads to implement these actions.

If the Forest Service Roadless Area Conservation Final Rule and Record of Decision are implemented, road construction and reconstruction and timber harvest, except for stewardship purposes, will be prohibited on 241,000 ha, or 31%, of the 77,000 ha of NFS lands.

4.3.4. Watershed-Riparian Relationships

Runoff and erosion processes are the key factors that affect the stability of riparian ecosystems, and their surrounding watersheds. By recognizing the relationship between watershed condition and riparian health[26], a watershed manager has a framework to improve the riparian health and surrounding watershed condition[27] (DeBano and Schmidt, 1989; LaFayette and DeBano, 1990; Baker et al., 1998; Hornbeck and Kochenderfer, 2000). Water flow and sediment movement into and through riparian ecosystems are controlled by the vegetation, physiology, geologic formations, and hydrology of these linked ecological-hydrologic systems. The response of these systems to disturbances, such as a timber harvesting, vegetative conversion, or fire, must be considered within this context.

4.3.4.1. Dynamic Equilibrium

A riparian ecosystem is considered healthy if channel sediment deposition (aggradation) and channel erosion (degradation) are at equilibrium (Heede, 1980; DeBano and Schmidt, 1989; Baker et al., 1998). A healthy riparian ecosystem maintains a dynamic equilibrium between the streamflow forces producing change and those resisting change to its vegetative, geomorphic, and structural features (Figure 4.2). The dynamic equilibrium is sufficiently stable in this condition and external disturbances on the surrounding hillslopes often occur without altering the equilibrium. However, when large changes in erosional and sediment-deposition occur, maintaining equilibrium is impossible.

The health and stability of riparian ecosystems are thrown out of equilibrium quicker and more permanently when degradation processes dominate the watershed-riparian system than during periods of deposition (Heede, 1980; Baker et al., 1998). Excessive channel incision can intercept and drain the site's water table, which is close to the surface in healthy

---

[26] The stage of vegetative, edaphic, geomorphic, and hydrologic development of the ecosystem and the degree of structural integrity of the ecosystem.

[27] The state of a watershed in relation to its vegetative cover, streamflow regime, sediment and nutrient outputs, and site productivity.

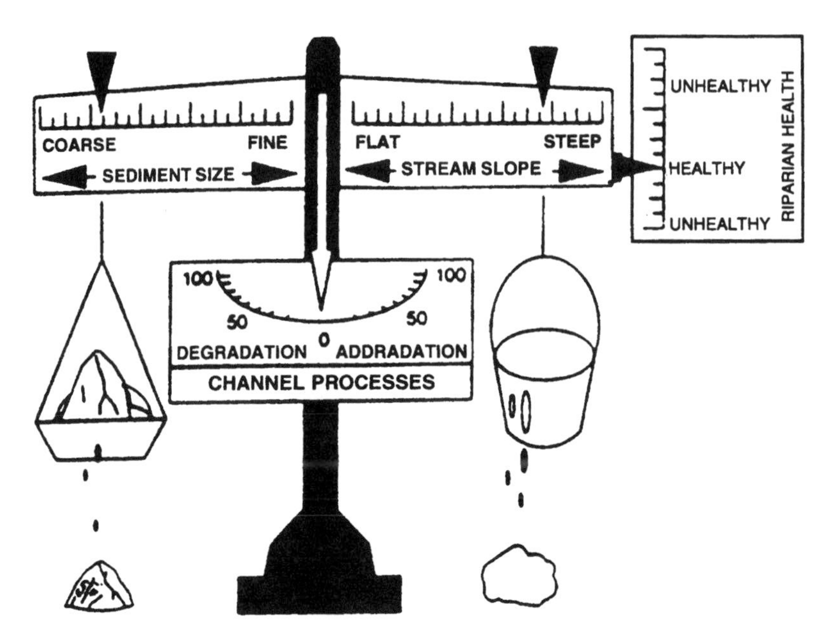

Figure 4.2. Balance of streamflow and sediment movement to maintain a dynamic equilibrium and support a healthy riparian ecosystem (adapted from Rosgen, 1980, based on Lane, 1955).

riparian ecosystems. Lowering the water table can rapidly desiccate the site, alter plant compositions, and reduce plant diversity, creating an unhealthy riparian ecosystem. When excessive deposition occurs because of a disturbance on the surrounding watersheds, channels can become braided and so shallow that they easily shift their locations, with resulting bank and channel erosion. Once the disturbance has been eliminated or otherwise mitigated, eliminating the excess sediment is usually easier for an aggraded system than it is for a degraded system to reestablish its original channel level.

4.3.4.2. Management Implications

Research efforts focusing on riparian ecosystem functioning and changing public attitudes have affected how these ecosystems and their surrounding watersheds are managed. Today, riparian ecosystems in the United States are not viewed as a major source of water loss or as a potential source of additional water obtained through the eradication of streamside vegetation, instead they are considered: (Baker et al., 1998; 1999; Verry et al., 2000a)

- Necessary habitats for a variety of wildlife species;
- Biological filters of sediment and unwanted organic matter that otherwise would enter and degrade stream systems;
- Areas for the controlled grazing of livestock and wildlife; and
- Unique sites for bird watching, camping, and picnicking.

Maintaining the integrity of riparian ecosystems requires careful management and, sometimes, protection from timber harvesting, livestock grazing, and recreational use.

While millions of dollars have been spent in the last century on riparian restoration and fish-habitat improvement, in many regions of the country, riparian ecosystems were disappearing and others were falling into serious degradation. Today, assessing the conditions of riparian ecosystems to determine which are unstable and are out of equilibrium and restoring riparian ecosystems that are not functioning properly are being emphasized (Baker et al., 1998, 1999; Verry et al., 2000a; 2000b).

4.3.5. Impacts of Urbanization

Urbanization causes major changes in watershed hydrology. The continually increasing urbanization in the United States has caused significant increases in the volume and peak of streamflows and pollutant loads to local water bodies. Depending on the region, impacts commonly associated with urban runoff (DeWalle et al., 2000; Drennan et al., 2000; Sotir et al., 2000) include:

- Flooding;
- Increased erosion and sediment deposition;
- Increased water temperature;
- Degradation of stream channels and riparian buffers;
- Losses of in-stream biota;
- Violations of ecological water-quality standards and human health; and
- Lowering aesthetics due to increased amounts of litter, turbidity, and runoff odor.

Urbanization is generally thought to reduce low flows due to a reduced groundwater recharge. However, while a reduction in low flows can occur on small watersheds immediately downstream from urban developments, larger watersheds could include return flows and flow-regulating influences from the urban developments. As urbanization continues, the impacts on the watershed hydrology of a region could become severe. There is increasing concern about how the urban environment and the subsequent impervious surfaces affect local hydrologic cycles (Box 4.4).

Management practices and regulatory actions with the greatest potential to improve water quality and riparian and stream habitats must be identified and incorporated into comprehensive watershed management programs encompassing urban development. Knowledge of the hydrologic functioning in the urban-wildland interface is incomplete. These and other issues will likely be the focus of urban-oriented watershed managers in the future.

Box 4.4 Hydrologic Impacts of Urbanization: An Example

Estimates are that less than 10% of the rain that fell on the watershed of the Los Angeles River turned into stormflow before 1930. This means that about 90% of the rainfall was available to infiltrate into the subsurface (Drennan et al., 2000). However, almost 50% of the rain that fell on this same watershed in the 1990s turned into stormflow. This indicates that the rapidly increasing expansion of the Los Angeles metropolitan area has greatly reduced the opportunity for rainfall to infiltrate into the subsurface and percolate to groundwater aquifers. An ever-increasing population with greater water demands, requires makes efforts to capture stormflow for recharge purposes a necessity. Greater attempts to capture stormflow are also necessary to mitigate the problems of increased sedimentation and decreased water quality.

4.3.6. Best Management Practices (BMPs)

The Clean Water Act of 1977 amended earlier federal water-quality legislation by increasing control of toxic pollutants and authorizing a grant program to help cover the costs to rural landowners of implementing BMPs (33 U.S.C. 1289) to control source pollution (Brown et al., 1993; Brooks et al., 1997; Beschta, 2000; Verry et al., 2000b). BMPs are applied before, during, and after pollution-producing activities to reduce or eliminate the introduction of pollutants into receiving waters. Many BMPs are known for forestry, livestock grazing, road construction, and agricultural production activities (Moore et al., 1979; Lynch et al., 1985; Chaney et al., 1990; LaFayette et al., 1992; Brooks et al., 1997). However, BMPs are not as well known for the control of some types of chemical pollutants.

Over the past decade, federal and state water-quality protection programs have improved, which suggests that many agencies are developing and implementing BMPs. General agreement exists that specifying BMPs is the most practical approach to meeting water-quality standards (Brown et al., 1993; Brooks et al., 1997; Beschta, 2000; Verry et al., 2000b). However, how these BMPs will be specified in the future is a complex matter, especially if the selected BMPs must be the most cost-effective. Specifying BMPs to attain designed water-quality standards in a cost-effective manner requires:

- An understanding of the cause-and-effect relationship between land disturbances on upland watersheds and downstream water quality and
- Knowing the costs of alternative approaches to mitigation.

An iterative process of BMP cost-effective use includes specifying, applying, monitoring, and correcting. Prescribing BMPs and using professional judgment will usually control source pollution within specified water-quality standards. Then reassessing the BMPs as new information becomes available is necessary.

4.4. RESEARCH NEEDS

Many research needs affect planning and management of natural resources on a watershed. Among these research needs are those identified by the participants at the conference on "Land Stewardship in the 21st Century: The Contributions of Watershed Management" including:

- Improving planning tools that concern the human elements of watershed management;
- Attaining a comprehensive understanding of basic hydrology and the effects of land-use disturbances on water and other natural resources on a landscape scale;
- Enhancing the capabilities of monitoring and evaluating techniques in assessing the effectiveness of watershed management programs;
- Encompassing coastal estuaries, lake shorelines, and streamside habitats in comprehensive watershed research programs;
- Improving the participatory planning of watershed management programs; and
- Developing mechanisms for evaluating the effectiveness of watershed policies and programs.

4.4.1. Planning Tools

Tools to deal with the human element of watershed management are necessary to incorporate stakeholders' ideas into the planning of responsive practices, projects, and programs. These planning tools should focus on how to change human behavior and how to communicate and work effectively with all concerned stakeholders (Adams, et al., 2000). Other tools, methods, and protocols are needed for abstract processes, such as developing ways to build a consensus for preserving natural resources and the quality of life and determining the effectiveness of land-use planning and controlled growth.

4.4.2. Hydrologic Research

Future investments in basic and applied hydrologic research are needed to improve the understanding of the relationship between land uses and the variables of interest at the watershed scale (Adams et al., 2000; Sidle, 2000; Thorud et al., 2000). For example, additional information is needed concerning the complex pathways traveled by subsurface water on steep hillslopes, and its interactions with soil strength, erosion processes, and landslide frequency. Watershed managers should trace the origins of pollution to individual response units and subbasins on watershed lands and identify the land-use practices causing the pollution. Effective incentives necessary to control this pollution must also be available.

Many early watershed-research efforts focused on hydrologic processes at a plot level or a small-watershed scale of less than a few hundred hectares. While such studies are still needed, it is also necessary to know how multiple land uses interact with each other to affect watershed resources on large scales (Sidle, 2000; Thorud et al., 2000). Watershed managers must be able to predict how disturbances, such as large-scale timber harvesting, livestock grazing, or fire, will affect water flows and other natural resources in large basins. Perhaps the most important disturbance of concern in the Western United States is fire (Pyne et al., 1996; DeBano et al., 1998). Increasing fuel loads from long-term fire-prevention policies on federal lands make this a critical issue.

4.4.3. Monitoring and Evaluation Techniques

A monitoring and evaluation program is prerequisite to the design and execution of watershed management programs because of incomplete or uncertain knowledge about the interrelationships among program activities, natural-resource systems, the environment, and program impacts on people. The program unfolds through monitoring and evaluation efforts, which helps improve a manager's ability to accomplish the watershed program and plan for program steps and future watershed management programs (Casley and Kumar, 1987; Brooks et al., 1990).

Unfortunately, any monitoring and evaluation program is incomplete and imperfect in obtaining information on the performance of a watershed management program. Information gained through monitoring will still contain uncertainties. Better systems for monitoring and evaluating streamflow regimes, water-quality constituents, and other variables of interest to managers, decision-makers, and the public are needed (Thorud et al., 2000). Additionally, improved methods of data archiving and processing and better visualization capabilities that make the data and information obtained through the monitoring and evaluation of watershed programs more accessible and useful are essential.

4.4.4. Coastal Estuaries, Lake Shorelines, and Streamside Habitats

Watershed research programs must encompass coastal estuaries, lake shorelines, and stream-sides that are critical fish habitats. A major driver for managing and regulating land-use practices on the watershed lands in the Western United States that drain into the Pacific Ocean is the restrictions associated with threatened or endangered runs of anadromous fish (Meehan, 1991; Stouder et al., 1997; Beschta, 2000; Mann and Plummer, 2000; Thorud et al., 2000). Throughout the United States and worldwide, the relationship of ocean and estuaries to other terrestrial ecosystems that affect fish populations must be better understood. A similar situation confronts watershed managers in freshwater environments. Watersheds surrounding lakes and streams must be protected from disturbances that lead to excessive erosion and high rates of sedimentations that detrimentally affect fish habitats. Therefore, increased knowledge of the relationships between land-use practices on watershed lands, the resulting erosion and sedimentation, and the quality of fish habitats is required (Gorman and Karr, 1978; Everest and Harr, 1982; Kohler and Hubert, 1993; Cole, 1994). Recognizing and implementing watershed management practices that sustain desirable habitat conditions for productive fish populations is essential.

4.4.5. Participatory Planning Process

Successful watershed planning that involves a variety of property owners, land managers, and natural-resource users in multiple land-use activities is rare. However, efforts are growing in response to the increasing scarcity of land, water, and other natural resources and legislative actions concerning endangered species. All groups of people must share needed information and effective decision-making tools, including those that have not been part of past planning processes. Research can bridge the organizational barriers to information exchange and prevent jurisdictional and institutional boundaries between and within the various levels of government and the private sector from hindering effective watershed management programs (Brooks et al., 1990; 1997; Ffolliott et al., 1995; Thorud et al., 2000). Commonly large blocks of federal land are juxtaposed with state and private land in a large watershed or river basin in many regions of the United States. Each of these ownerships has different policies, regulations, and limitations on the role of public participation in the planning and decision-making process. These differences should be coordinated and sharing science and knowledge through improved technology transfer will help to span these boundaries.

4.4.6. Policy and Program Evaluation

More research is needed to develop mechanisms for evaluating the effectiveness of watershed policies and programs. These mechanisms must include a more specific measure of the effectiveness and success of a policy or program than the simplistic enumerations of outcomes, for example, the number of species listed or the number of enforcement notices issued. Current evaluation efforts often result in a determination that the policies or programs have failed or been ineffective because how to comprehensively evaluate the impacts of the policies or programs is unknown (Brooks et al., 1994; Quinn et al., 1995; Cortner and Moote, 1999; Brooks and Eckman, 2000; Eckman et al., 2000; Thorud et al., 2000).

If adaptive management principles are applied in watershed management—as many planners, managers, and decision-makers propose—a new generation of evaluation mechanisms must be developed and implemented. Without these evaluation capabilities, the promise of adaptive management as a systematic and rigorous approach for evaluating the outcomes of a watershed management approach to land stewardship will not be fully realized. The reliable feedback that adaptive management is intended to provide is crucial to improving subsequent actions and accommodating change.

4.5. SUMMARY

Society must confront many issues to implement effective watershed management practices, projects, and programs. On a global scale, these issues include increasing water scarcity, water pollution, effects of droughts, floods, and torrents, landslides, hydrologic response of headwater areas, cumulative watershed effects, scarcity of land and other natural resources, and land and resource tenure. Sustaining high-quality water that is readily available in adequate quantities for human use will increasingly become a dominant issue in the United States. Other issues identified as high priorities in the United States are the maintenance of water-quality standards, role of prescribed fire, affect of roads and trails, watershed-riparian relationships, impacts of urbanization, and implementation of BMPs.

Many of these issues concern the status and future of the watershed management planning process. This planning process is complex, however, and often difficult to follow and appreciate because of the physical, biological, and social interactions that are the foundation of watershed management. Nevertheless, people have a responsibility to act together in society to conserve natural resources and preserve their integrity for future generations. The outcome of this participatory effort is land stewardship based on sustainability in the use of natural resources through watershed management.

REFERENCES

Abbe, T. B., Montgomery, D. R. Large woody debris jams, channel hydraulics and habitat formation in large rivers. Regulated Rivers 1996; 12:201-221.

Adams, C., Noonan, T., Newton, B. "Watershed Management in the 21st Century: National Perspectives." In *Land Stewardship in the 21st Century: The Contributions of Watershed Management,* P. F. Ffolliott, M. B. Baker, Jr., C. B. Edminister, M. C. Dillon, and K. L. Mora, tech. coords. Fort Collins, CO: Rocky Mountain Research Station, USDA Forest Service, Proceedings RMRS-P-13, 2000, pp. 21-29.

Anaya-Garduno, M. A., Pascual-Moncayo, M. A., Zarate-Zarate, R., eds., *Sustainable Development for Our Common Future.* Mexico City, Mexico: International Desert Development Commission, 1994.

Anderson, H. W., Hoover, M. D., Reinhart, K. G., *Forest and Water: Effects of Forest Managment on Floods, Sedimentation, and Water Supply.* Berkeley, CA: Pacific Southwest Forest and Range Experiment Station, USDA Forest Service, General Technical Report PSW-18, 1976.

Aplet, G. H., Johnson, N., Olson, J. T., Sample, V. A., eds., *Defining Sustainable Forestry.* Covello, CA: Island Press, 1993.

Armitage, F. B., *Irrigated Forestry in Arid and Semi-Arid Lands: A Synthesis.* Ottawa, Canada: International Development Research Center, 1987.

Ayers, R. S., Westcott, D. W., *Water Quality for Irrigation.* Rome: Food and Agriculture Organization of the United Nations, FAO Irrigation and Drainage Paper 29, 1976.

Baker, M. B., Jr. "Hydrologic and Water Quality Effects of Fire." In *Effects of Fire Management of Southwestern Natural Resources,* J. S. Krammes, ed. Fort Collins CO: Rocky Mountain Forest and Range Experiment Station, USDA Forest Service, General Technical Report RM-191, 1990, pp. 31-42.

Baker, M. B., Jr., DeBano, L. F., Ffolliott, P. F. "Changing Values of Riparian Ecosystems." In *History of Watershed Research in the Central Arizona Highlands,* M. B. Baker, Jr., compiler. Fort Collins, CO: Rocky Mountain Research Station, USDA Forest Service, General Technical Report RMRS-GTR-29, 1999, pp. 43-47.

Baker, M. B., Jr., DeBano, L. F., Ffolliott, P. F., Gottfried, G. J. "Riparian-Watershed Linkages in the Southwest." In *Rangeland Management and Water Resources: Proceedings of the American Water Resources Association Specialty Conference,* D. E. Potts, ed. Herndon, VA: American Water Resources Association, TPS-98-1, 1998, pp. 347-357.

Becker, A., Braun, P. Disaggregation, aggregation and spatial scaling in hydrologic modeling. Journal of Hydrology 1999; 217:239-252.

Bell, J. W. "The National Forest Road System: A Public Policy Issue for the 21st Century." In *Watershed 2000: Science and Engineering Technology for the New Millennium,* M. Flug and D. Frevert, eds. Reston, VA: American Society of Civil Engineers, 2000. [CD-ROM] Windows.

Beschta, R. L. "Watershed Management in the Pacific Northwest: The Historical Legacy." In *Land Stewardship in the 21st Century: The Contributions of Watershed Management,* P. F. Ffolliott, M. B. Baker, Jr., C. B. Edminster, M. C. Dillon, and K. L. Mora, tech. coords. Fort Collins, CO: Rocky Mountain Research Station, USDA Forest Service, Proceedings RMRS-P-13, 2000, pp. 109-116.

Bishop, D. M., Stevens, M. E., *Landslides on Logged Areas, Southeast Alaska.* Fairbanks, AK: Northern Forest Research Station, USDA Forest Service, Research Report NOR-1, 1964.

Boyer, D. M., Perry, H. D. Fecal coliform concentrations in runoff from a grazing, reclaimed surface mine. Water Resources Bulletin 1987; 23:911-917.

Brooks, K. N., Eckman, K. "Global Perspective of Watershed Management. In *Land Stewardship in the 21st Century: The Contributions of Watershed Management,* P. F. Ffolliott, M. B. Baker, Jr., C. B. edminister, M. C. Dillon, and K. L. Mora, tech. coords. Fort Collins, CO: Rocky Mountain Research Station, USDA Forest Service, Proceedings RMRS-P-13, 2000, pp. 11-20.

Brooks, K. N., Ffolliott, P. F., Gregersen, H. M., DeBano, L. F., *Hydrology and the Management of Watersheds.* Ames, IO: Iowa State University Press, 1997.

Brooks, K. N., FFolliott, P. F., Gregersen, H. M., Easter, K. W., *Policies for Sustainable Development: The Role of Watershed Management.* Washington, DC: U.S. Department of State, EPAT Policy Brief 6, 1994.

Brooks, K. N., Gregersen, H. M., Ffolliott, P. F., Tejwani, K. G. "Watershed Management: A Key to Sustainability." In *Managing the World's Forests: Looking for Balance Between Conservation and Development,* N. P. Sharma, ed. Dubuque, IA: Kendall/Hunt Publishing Company, 1992, pp. 455-487.

Brooks, K. N., Gregersen H. M., Lundgren, A. L., Quinn, R. M., *Manual on Watershed Management Project Planning, Monitoring and Evaluation.* College, Laguna, Philippines: ASEAN-US Watershed Project, 1990.

Brown, R. J., Hanson, M. E., Liverman, D. M., Merideth, R. W., Jr. Global sustainability: Toward definition. Environmental Management 1987; 11:713-719.

Brown, T. C., Brown, D., Brinkley, D. Laws and programs for controlling non-point source pollution in forest areas. Water Resources Bulletin 1993; 23:1149-1160.

Brown, V. A., McDonnell, J. J., Burns, D. A., Kendall, C. The role of event water, a rapid shallow flow component, and catchment size in summer stormflow. Journal of Hydrology 1999; 217:171-190.

Burgess, S. J., Wigmosta, M. S., Menna, J. M. Hydrological effects of land-use change in a zero-order catchment. Journal of Hydrologic Engineering 1998; 3:86-97.

Burroughs, E. R., Jr., King, J. G., *Reduction of Soil Erosion on Forest Roads.* Ogden, UT: Intermountain Research Station, USDA Forest Service, General Technical Report INT-264, 1989.

Casley, D. J., Kumar, K., *Project Monitoring and Evaluation in Agriculture.* Baltimore, MD: The Johns Hopkins University Press, 1987.

Chandra, S., Bhatia, K. K. S. "Water and Watershed Management in India: Policy Issues and Priority Areas for Future Research." In *Land Stewardship in the 21st Century: The Contributions of Watershed Management,* P. F. Ffolliott, M. B. Baker, Jr., C. B. Edminster, M. C. Dillon, and K. L. Mora, tech. coords. Fort Collins, CO: Rocky Mountain Research Station, USDA Forest Service, Proceedings RMRS-P-13, 2000, pp. 158-165.

Chaney, E., Elmore, W., Platts, W. S., *Livestock Grazing on Western Riparian Areas.* Washington, DC: U.S. Environmental Protection Agency, 1990.

Coghlan, G. and R. Sowa, [unpublished draft]. National Forest Road System and Use. Washington DC: United States Department of Agriculture Forest Service, Engineering Staff, Washington Office.1998.

Cole, G. A., *Textbook of Limnology*. Prespect Heights, IL: Waveland Press, 1994.

Conway, G. Agroecosystem analysis. Agricultural Administration 1985; 20:31-55.

Copley, T. L., Forrest, L. A., McColl, A. G., Bell, F. G., *Investigations in Erosion Control and Reclamation of Eroded Land in the Central Piedmont Conservation Experiment Station*. Washington, DC: U.S. Department of Agriculture, USDA Technical Bulletin 873, 1944.

Cortner, H. J., Moote, M. A., *The Politics of Ecosystem Management*. Covelo, CA: Island Press, 1999.

Cortner, H. J., Zwolinski, M. J., Carpenter, E. H., Taylor, J. G. Public support for fire-management policies. Journal of Forestry 1984; 82:359-361.

Czech, B., Ffolliott, P. F. "The Let-Burn Policy: Implications in the Madrean Province of the Southwestern United States." In *Effects of Fire on Madrean Province Ecosystems*, P. F. Ffolliott, L. F. DeBano, M. B. Baker, Jr., G. J. Gottfried, G. Solis-Garza, C. B, Edminster, D. G. Neary, L. S. Allen, L. S., and R. H. Hamre, tech. coords. Fort Collins, CO: Rocky Mountain Forest and Range Experiment Station, USDA Forest Service, General Technical Report RM-GTR-289, 1996, pp. 233-225.

DeBano, L. F., Schmidt, L. J., *Improving Southwestern Riparian Areas Through Watershed Management*. Fort Collins, CO: Rocky Mountain Forest and Range Experiment Station, USDA Forest Service, General Technical Report RM-182, 1989.

DeBano, L. F., Baker, M. B., Jr., Ffolliott, P. F. Evaluating the effects of fire on dissolved constituents in streamflow: A commentary. Hydrology and Water Resources in Arizona and the Southwest 1997; 27:33-38.

DeBano, L. F., Neary, D. G., Ffolliott, P. F., *Fire's Effects on Ecosystems*. New York: John Wiley & Sons, Inc., 1998.

de Steiguer, J. E. "Sustaining Flows of Crucial Watershed Resources." In *Land Stewardship in the 21st Century: The Contributions of Watershed Management,* P. F. Ffolliott, M. B. Baker, Jr., C. B. Edminster, M. C. Dillon, and K. L. Mora, tech. coords. Fort Collins, CO: Rocky Mountain Research Station, USDA Forest Service, Proceedings RMRS-P-13, 2000, pp. 215-220.

DeWalle, D. R., Swistock, B. R., Johnson, T. E. "Streamflow Variations with Populations Growth on Urbanizing Catchments in the United States." In *Watershed 2000: Science and Engineering Technology for the New Millennium*, M. Flug and D. Frevert, eds. Reston, VA: American Society of Civil Engineers, 2000. [CD-ROM] Windows.

Drennan, M., Moore, G., Lipkis, A., Davis, S. "How the Impacts of Urbanization are being Addressed in the Los Angeles River Watersheds." In *Watershed 2000: Science and Engineering Technology for the New Millennium*, M. Flug and D. Frevert, eds. Reston, VA: American Society of Civil Engineers, 2000. [CD-ROM] Windows.

Dunne, T., Leopold, L. B., *Water in Environmental Planning*. San Francisco, CA: W. H. Freeman and Company, 1978.

Eckman, K., Gregersen, H. M., Lundgren, A. L. "Watershed Management and Sustainable Lessons Learned and Future Directions." In *Land Stewardship in the 21st Century: The Contributions of Watershed Management,* P. F. Ffolliott, M. B. Baker, Jr., C. B. Edminster, M. C. Dillon, and K. L. Mora, tech. coords. Fort Collins, CO: Rocky Mountain Research Station, USDA Forest Service, Proceedings RMRS-P-13, 2000, pp. 37-43.

Engelmann, R., LeRoy, P., *Sustaining Water: Population and the Future of Renewable Water Supplies*. Washington, DC: Population Action International, 1993.

Engelman, R., LeRoy, P., *Conserving Land: Population and Sustainable Food Production*. Washington, DC: Population Action International, 1995.

Everest, F. H., Harr, D. R., *Influence of Forest Rangeland Management on Anadromous Fish Habitat in Western North America Silvicultural Treatments*. Portland, OR: Pacific Northwest Forest and Range Experiment Station, USDA Forest Service, General Technical Report PNW-134, 1982.s

Ffolliott, P. F., Brooks, K. N., Gregersen, H. M., Lundgren, A. L., *Dryland Forestry: Planning and Management*. New York: John Wiley & Sons, Inc., 1995.

Ffolliott, P. F., Baker, M. B., Jr., Edminster, C. B. Dillon, M. C., Mora, K. L., tech. coords. *Land Stewardship in the 21st Century: The Contributions of Watershed Management*. Fort Collins, CO: Rocky Mountain Research Station, USDA Forest Service, Proceedings RMRS-P-13, 2000.

Foster, I., Gurnell, A., Webb, B., eds., *Sediment and Water Quality in River Catchments*. Chichester, NY: John Wiley & Sons, Inc., 1995.

Furniss, M. J. "Road Analysis: Informing Decisions About Managing the National Forest Transportation System." In *Watershed 2000: Science and Engineering Technology for the New Millennium*, M. Flug and D. Frevert, eds. Reston, VA: American Society of Civil Engineers, 2000. [CD-ROM] Windows.

Gardner-Outlaw, T., Engelman, R., *Sustaining Water, Easing Scarcity: A Second Update*. Washington, DC: Population Action International, 1997.

Gardner-Outlaw, T., Engelman, R., *Forest Futures: Population, Consumption and Wood Resources*. Washington, DC: Population Action International, 1999.

Gelt, J. "Watershed Management: A Concept Evolving to Meet New Needs." In *Land Stewardship in the 21st Century: The Contributions of Watershed Management,* P. F. Ffolliott, M. B. Baker, Jr., C. B. Edminster, M. C. Dillon, and K. L. Mora, tech. coords. Fort Collins, CO: Rocky Mountain Research Station, USDA Forest Service, Proceedings RMRS-P-13, 2000, pp. 65-73.

Gorman, O. F., Karr, J. R. Habitat structure and stream fish communities. Ecology 1978; 59:507-515.

Gottfried, G. J., DeBano, L. F. "Streamflow and Water Quality Responses to Preharvest Prescribed Burning in an Undisturbed Ponderosa Pine Watershed." In *Effects of Fire Management of Southwestern Natural Resources,* J. S. Krammes, ed. Fort Collins CO: Rocky Mountain Forest and Range Experiment Station, USDA Forest Service, General Technical Report RM-191, 1990, pp. 222-228.

Gregersen, H. M., Lundgren, A. L., *Forestry for Sustainable Development: Concepts and a Framework for Action*. St. Paul, MN: Forestry for Sustainable Development Program, College of Natural Resources, University of Minnesota, Working Paper 1, 1990.

Gregersen, H. M., Easter, W. K., de Steiguer, J. E. "Responding to Increased Needs and Demands for Water." In *Land Stewardship in the 21st Century: The Contributions of Watershed Management,* P. F. Ffolliott, M. B. Baker, Jr., C. B. Edminster, M. C. Dillon, and K. L. Mora, tech. coords. Fort Collins, CO: Rocky Mountain Research Station, USDA Forest Service, Proceedings RMRS-P-13, 2000, pp. 238-246.

Gregersen, H., Lundgren, A., Byron, N. Forestry for sustainable development: Making it happen. Journal of Forestry 1998; 96(3):6-10.

Gregersen, H. M., Brooks, K. N., Dixon, J. A., Hamilton, L. S., *Guidelines for Economic Appraisal of Watershed Management Projects*. Rome: Food and Agriculture Organization of the United Nations, FAO Conservation Guide 17, 1987.

Grier, C. C. Wildfire effects on nutrient distribution and leaching in a coniferous ecosystem. Canadian Journal of Forestry 1975; 5:559-607.

Heede, B. H., *Stream Dynamics: An Overview for Land Managers*. Fort Collins, CO: Rocky Mountain Forest and Range Experiment Station, USDA Forest Service, General Technical Report RM-72, 1980.

Hewlett, J. D., Hibbert, A. R. "Factors Affecting the Response of Small Watersheds to Precipitation in Humid Areas." In *Forest Hydrology: Proceedings of a National Science Foundation Advanced Symposium*, W. E. Sopper and H. W. Lull, eds. New York: Pergamon Press, 1967, pp. 275-290.

Hornbeck, J. W., Kochenderfer, J. N. "Linkages Between Forests and Streams: A Perspective in Time." In *Riparian Management in Forests of the Continental Eastern United States*, E. S. Verry, Hornbeck, J. W., Dolloff, C. A., eds. Boca Raton, FL: Lewis Publishers, 2000, pp. 89-98.

Howes, D. E., Kenk, E., *Terrain Classification System for British Columbia*. Victoria, BC, Canada: Ministry of Crown Lands, Ministry of Environment Manual 10, 1988.

Jones, J. A., Grant, G. E. Peak flow responses to clearcutting and roads in small and large basins, western Cascades. Water Resources Research 1996; 32:959-974.

Kauffman, J. B., Kreuger, W. C. Livestock impacts riparian ecosystems and streamside management implications: A review. Journal of Range Management 1984; 37:430-438.

Keeney, D. R., Alexander, R. H. Droughts, floods and sustainability. Water Resources Update 1994; 94-15-20.

Kohler, C. C., Hubert, W. A., eds., *Inland Fisheries Management in North America*. Bethesda, MD: American Fisheries Society, 1993.

LaFayette, R. A., DeBano, L. F. "Watershed Condition and Riparian Health: Linkages." In *Proceedings of the symposium sponsored by the Committee on Watershed Management of the Irrigation and Drainage Division of the American Society of Civil Engineers in conjunction with the SCE Irrigation and Drainage Division Conference*, R.E. Riggins, E.B. Jones, R. Singh, and P.A. Rechard, eds. Reston, VA: American Society of Civil Engineers, 1990, pp. 473-484.

LaFayette, R. A., Pruit, J. R., Zeedyk, W. D., *Riparian Area Enhancement Through Road Management*. Albuquerque, NM: Southwestern Region, USDA Forest Service, 1992.

Lal, R. "Soils of the Tropics and Their Management for Plantation Forestry." In *Management of Soil, Nutrients, and Water in Tropical Plantation Forests*, E. K. S. Nambiar and A. G. Brown, eds. Sidney, Australia: Australian Centre for International Agriculture Research , 1997, pp. 97-123.

Lamb, J. C., *Water Quality and Its Control*. New York: John Wiley & Sons, Inc., 1985.

Lane, E. W. "The Importance of Fluvial Hydrology in Hydraulic Engineering." In *Proceedings of the American Society of Civil Engineers*, 1955: 81(745):117.

Leopold, L. Flood hydrology and the floodplain. Water Resources Update 1994; 94:11-14.
Loftis, J. C., MacDonald, L. H. Exploring the benefits of paired watersheds for detecting cumulative effects. In *Watershed 2000: Science and Engineering Technology for the New Millennium*, M. Flug and D. Frevert, eds. Reston, VA: American Society of Civil Engineers, 2000. [CD-ROM] Windows.
Lynch, J. A., Corbett, E. S., Mussallem, K. Best Management Practices for controlling non-point-source pollution on forested watersheds. Journal of Soil and Water Conservation 1985; 40:164-167.
Lyons, J. K., Beschta, R. L. Land use, floods, and channel changes: Upper Middle Fork Willamette River, Oregon (1936-1980). Water Resources Research 1983; 19:463-471.
MacDonald, L. H. "Predicting and Managing Cumulative Watershed Effects." In *Watershed 2000: Science and Engineering Technology for the New Millennium*, M. Flug and D. Frevert, eds. Reston, VA: American Society of Civil Engineers, 2000. [CD-ROM] Windows.
Mann, C. C., Plummer, M. L. Can science rescue salmon? Science 2000; 289:716-719.
Maser, C., *Sustainable Forestry: Philosophy, Science, and Economics*. Delray Beach, FL: St. Lucie Press, 1994.
McGee, J. "Getting to Our Public Lands." In *Sunday Opinion*. Tucson, AZ: Arizona Daily Star (August 27) 2000, Section B, p. 7.
Meehan, W. R., *Influences of Forest and Rangeland Management on Salmonid Fishes and Their Habitats*. Bethesda, MD: American Fisheries Society, 1991.
Megahan, W. F., "Reducing Erosional Impacts of Roads. In *Guidelines for Watershed Management*, S. H. Kunckle and J. L. Thames, eds. Rome: Food and Agricultural Organization of the United Nations, 1977, pp.237-261.
Megahan, W. F., Hornbeck, J. "Lessons Learned in Watershed Management: A Retrospective View." In *Land Stewardship in the 21st Century: The Contributions of Watershed Management,* P. F. Ffolliott, M. B. Baker, Jr., C. B. Edminster, M. C. Dillon, and K. L. Mora, tech. coords. Fort Collins, CO: Rocky Mountain Research Station, USDA Forest Service, Proceedings RMRS-P-13, 2000, pp. 177-188.
Megahan, W. F., Kidd, W. J. Effects of logging and logging roads on erosion and sediment deposition. Journal of Forestry 1972; 70:136-141.
Megahan, W. F., Day, N. F., Bliss, T. M. Landslide occurrence in the western and central northern Rocky Mountain physiographic province in Idaho. Proceedings of the 5th North American Forest Soils Conference; C.T. Youngberg, ed. ; Fort Collins, CO: Colorado State University, 1978, pp. 115-139.
Metz, A. Mideast's lifeblood: A region's water crisis - from droughts to dams to drying rivers, Mideast is steeped in crisis. Newsday.com, (June 25) 2000, pp. 1-7.
Moore, E., James, E., Kinsinger, F., Pitney, K., Sainsbury, J., *Livestock Grazing Management and Water Quality Protection*. Washington, DC: U.S. Environmental Protection Agency, EPA 910/9-79-67, 1979.
Mouat, D. A., Hutchinson, C. F., *Desertification in Developed Countries*. Dordrecht, The Netherlands: Kluwer Academic Press, 1995.
Murphy, M. L., Koski, K. V. Input and depletion of woody debris in Alaska streams and implications for streamside management. North American Journal of Fisheries Management 1989; 9:427-436.
Nielsen, T. H., Wright, R. H., Vlasic, T. C., Spangle, W. E., *Relative Slope Stability and Land-Use Planning in the San Francisco Bay region, California*. Washington, DC: U.S. Geological Survey, Professional Paper 944, 1979.
O'Loughlin, C. L., Pearce, A. J. Influence of cenozoic geology on mass movement and sediment yield in response to forest removal, North Westland, New Zealand. Bulletin of the International Association of Engineering Geologists 1976; 14:41-46.
Pimental, D., Harvey, C., Resosudarmo, P., Sinclair, K., Kurtz, D., McNair, M., Crist, S., Shpritz, L., Fritton, L., Saffouri, R., Blair, R. Environmental and economic costs of soil erosion and conservation benefits. Science 1995; 267:1117-1122.
Postel, S., *Last Oasis: Facing Water Scarcity*. New York: W. W. Norton Company, 1992.
Potts, D. E., ed., *Rangeland Management and Water Resources: Proceedings of the American Water Resources Association Specialty Conference*. Herndon, VA: American Water Resources Association, TPS-98-1, 1998.
Prosser, I. P., Soufi, M. Controls on gully formation following forest clearing in a humid temperate environment. Water Resources Research 1998; 34:3661-3671.
Pyne, S. J., Andrews, P. L., Laven, R. D., *Introduction to Wildland Fire*. New York: John Wiley & Sons, Inc., 1996.
Quinn, R. M., Brooks, K. N., Ffolliott, P. F., Gregersen, H. M., Lundgren, A. L., *Reducing Resource Degradation: Designing Policy for Effective Watershed Management*. Washington, DC: U.S. Department of State, EPAT Working Paper 22, 1995.
Reid, L. M., *Research and Cumulative Watershed Management Effects*. Berkeley, CA: Pacific Southwest Forest and Range Experiment Station, USDA Forest Service, General Technical Report PSW-GTR-141, 1993.

Repetto, R., ed., *The Global Possible: Resources, Development, and the New Century*. New Haven, CT: Yale University Press, 1985.

Rice, R. M., Corbett, E. S., Bailey, R. G. Soil slips related to vegetation, topography, and soil in southern California. Water Resources Research 1969; 5:647-659.

Robichaud, P. R., Waldrop, T. A. A comparison of surface runoff and sediment yields from low- and high-severity site preparation burns. Water Resources Bulletin 1994; 30:27-34.

Rosegrant, M. W., *Water Resources in the Twenty-First Century: Challenges and Implications for Action*. Washington, DC: International Food Policy Research Institute, Food, Agriculture, and the Environment Discussion Paper 20, 1997.

Rosgen, D. L. "Total Potential Sediment." In *An Approach to Water Resources Evaluation, Non-point Silvicultural Sources*. Athens, GA: U.S. Environmental Protection Agency, Environmental Research Laboratory, 1980, pp. VI-1-VI-43.

Satterlund, D. R., Adams, P. W., *Wildland Watershed Management*. New York: John Wiley & Sons, Inc., 1992.

Scherr, S. J., Yadav, S., *Land Degradation in the Developing World: Implications for Food, Agriculture, and the Environment in 2020*. Washington, DC: International Food Policy Research Institute, Food, Agriculture, and the Environment Discussion Paper 14, 1996.

Schlesinger, W. H., Reynolds, J. F., Cunningham, G. L. Huenneke, L. F., Jarrell, W. M., Virginia, R. A., Whitford, W. G. Biological feedbacks in global desertification. Science 1990; 247:1043-1048.

Scott, D. F. The hydrological effect of fire in South Africa mountain catchments. Journal of Hydrology 1993; 150: 409-432.

Sidle, R. C. A conceptual model of changes in root cohesion in response to vegetation management. Journal of Environmental Quality 1991; 20:43-52.

Sidle, R. C. "Watershed Challenges for the 21st Century: A Global Perspective for Mountainous Terrain." In *Land Stewardship in the 21st Century: The Contributions of Watershed Management,* P. F. Ffolliott, M. B. Baker, Jr., C. B. Edminster, M. C. Dillon, and K. L. Mora, tech. coords. Fort Collins, CO: Rocky Mountain Research Station, USDA Forest Service, Proceedings RMRS-P-13, 2000, pp. 45-56.

Sidle, R. C., Amacher, M. C. "Effects of Mining, Grazing, and Roads on Sediment and Water quality in Birch Creek, Nevada." In *Watershed Planning and Analysis in Action: Proceedings of the Symposium*, R. E. Riggins, E. B. Jones, R. Singh, and P. A. Rechard, eds. Reston, VA: American Society of Civil Engineers, 1990, pp. 463-472.

Sidle, R. C., Hornbeck, J. W. Cumulative effects: A broader approach to water quality research. Journal of Soil and Water Conservation 1991; 46:268-271.

Sidle, R. C., Sharma, A. Stream channel changes associated with mining and grazing in the Great Basin. Journal of Environmental Quality 1996; 25:1111-1121.

Sidle, R. C., Pearce, A. J., O'Loughlin, C. L., *Hillslope Stability and Land Use*. Washington, DC: American Geophysical Union, Water Resources Monograph 11, 1985.

Sidle, R. C., Tsuboyama, Y., Noguchi, S., Hosoda, I., Fujieda, M., Shimizu, T. Seasonal hydrologic response at various scales in a small forested catchment, Hitachi Ohta, Japan. Journal of Hydrology 1995; 168:227-250.

Sinowski,, W., Auerswald, K. Using relief parameters in a disciminant analysis to stratify grological areas with different spatial variability of soil properties. Geoderma 1999; 89: 113-128.

Sotir, R. B., Brosnan, T., Ferguson, B. K., Iosco, R., Moll, G., Schueler, T., Watson, R. "A Comprehensive Watershed Program for Atlanta Streams." In *Watershed 2000: Science and Engineering Technology for the New Millennium*, M. Flug and D. Frevert, eds. Reston, VA: American Society of Civil Engineers, 2000. [CD-ROM] Windows.

Squires, V. R., Sidahmed, A. E., eds., *Drylands: Sustainable Use of Rangelands into the Twenty-First Century*. Rome, Italy: International Fund for Agricultural Development, 1998.

Steeby, L. R. "Buffer Strips - Considerations in the Decision to Leave." In *Forest Land Uses and Stream Environment*, J. T. Krygier and J. D. Hall, eds. Corvallis, OR: Oregon State University, 1971, pp. 194-198.

Stouder, D. J., Bisson, P. A., Naiman, eds., *Pacific Salmon and Their Ecosystems: Status and Future Options*. New York: Chapman & Hall, 1997.

Stumnn, W., Morgan, J. J., *Aquatic Chemistry*. New York: John Wiley & Sons, 1995.

Swanson, F. J., Dryness, C. T. Impact of clearcutting and road construction on soil erosion by landslides in the western Cascades. Geology 1975; 3:393-396.

Taylor, J. G., Cortner, H. J., Gardner, P. D., Daniel, T. C., Zwolinski, M. J., Carptender, E. H. Recreation and fire management: Public concerns, attitudes, and perceptions. Leisure Sciences 1986; 8:167-187.

Tiedemann, A. R., Conrad, C. E., Dietrich, J. H., Hornbeck, J. W., Megahan, W. F., Viereck, L. A., Wade, D. D., *Effects of Fire on Water: A State-of-Knowledge Review*. Washington, DC: Washington Office, USDA Forest Service, General Technical Report WO-10, 1979.

The World Bank, *Desertification in the Sahelian and Sudanian Zones of West Africa*. Washington, DC: The World Bank, 1985.

The World Bank, *Water Resources Management*. Washington, DC: The World Bank, 1993.

Thomas, R. B., Megahan, W. F. Peak flow responses to clear-cutting and roads in small and large basins, western Cascades, Oregon: A second opinion. Water Resources Research 1998; 34:3393-3403.

Thomas, R., Colby, M., English, R., Jobin, W., Rassas, B., Reiss, P., *Water Resources Policy and Planning: Toward Environmental Sustainability*. Washington, DC: Irrigation Support Project for Asia and the Near East, Bureau for the Near East, US Agency for International Development, 1993.

Thorud, D. B., Brown, G. W., Boyle, B. J., Ryan, C. M. "Watershed Management in the United States in the 21st Century." In *Land Stewardship in the 21st Century: The Contributions of Watershed Management,* P. F. Ffolliott, M. B. Baker, Jr., C. B. Edminster, M. C. Dillon, and K. L. Mora, tech. coords. Fort Collins, CO: Rocky Mountain Research Station, USDA Forest Service, Proceedings RMRS-P-13, 2000, pp. 57-64.

Towns, E. S. "Resource Integration and Shared Outcomes at the Watershed Scale." In *Land Stewardship in the 21st Century: The Contributions of Watershed Management,* P. F. Ffolliott, M. B. Baker, Jr., C. B. Edminster, M. C. Dillon, and K. L. Mora, tech. coords. Fort Collins, CO: Rocky Mountain Research Station, USDA Forest Service, Proceedings RMRS-P-13, 2000, pp. 74-80.

Tracy, J. C., Bernknopf, R., Forney, W., Hill, K., A prototype for understanding the effects of TMDL standards: Values to sediment loads in the Lake Tahoe Basin. In *Watershed 2000: Science and Engineering Technology for the New Millennium*, M. Flug and D. Frevert, eds. Reston, VA: American Society of Civil Engineers, 2000. [CD-ROM] Windows.

U.S. Agency for International Development, *Water Resources Action Plan for the Near East*. Washington, DC: Irrigation Support Project for Asia and the Near East, Bureau for the Near East, U.S. Agency for International Development, 1993.

USDA Forest Service, *Roadless Area Conservation: Final Environmental Impact Statement*. Washington, DC: United States Department of Agriculture Forest Service, 2000.

USDA Forest Service, *Natural Resource Agenda*. Internet WWW page at URL http://www.fs.fed.us/news/agenda/nr30298.html, 1998.

USDA Forest Service, *Report of the Forest Service for Fiscal Year 1998*. Washington, DC: United States Department of Agriculture Forest Service, 1999.

Verry, E. S., Hornbeck, J. W., Dolloff, A. C., eds., *Riparian Management in Forests of the Continental Eastern United States*. Boca Raton, FL: Lewis Publishers, 2000a.

Verry, E. S., Hornbeck, J. W., Todd, A. H. "Watershed Research and Management in the Lake States and Northeastern United States." In *Land Stewardship in the 21st Century: The Contributions of Watershed Management,* P. F. Ffolliott, M. B. Baker, Jr., C. B. Edminster, M. C. Dillon, and K. L. Mora, tech. coords. Fort Collins, CO: Rocky Mountain Research Station, USDA Forest Service, Proceedings RMRS-P-13, 2000b, pp. 81-92.

Warren, S. D., Blackburn, W. H., Taylor, C. A. Effects of season and stage of rotation cycle on hydrologic condition of rangeland under intensive rotation grazing. Journal of Range Management 1986; 39:486-491.

Wright, H. A., Churchill, F. M., Stevens, W. C. Effect of prescribed burning on sediment, water yield, and water quality from dozed juniper lands in central Texas. Journal of Range Management 1976; 29:294-298.

Yaron, B., Daufors, E., Vaadia, Y., eds., *Arid Zone Irrigation*. New York: Springer-Verlag, 1973.

Zwolinski, M. J. "The Role of Fire in Management of Watershed Responses." In *Land Stewardship in the 21st Century: The Contributions of Watershed Management,* P. F. Ffolliott, M. B. Baker, Jr., C. B. Edminster, M. C. Dillon, and K. L. Mora, tech. coords. Fort Collins, CO: Rocky Mountain Research Station, USDA Forest Service, Proceedings RMRS-P-13, 2000, pp. 367-370.

5

WATERSHED MANAGEMENT CONTRIBUTIONS TO LAND STEWARDSHIP

Effective watershed management practices contribute to land stewardship by sustaining the physical and economic flows of crucial natural and economic resources. These resources include tangible commodities, such as water, timber, and forage for livestock production and intangible values, such as scenic quality, experiencing solitude, and existence and bequest values. Securing clean water and maintaining the health and stability of sensitive watershed-riparian systems by recognizing the complexity of responses in these ecosystems will continue to be significant watershed management contributions to land stewardship. Enhancing opportunities for investment, employment, and income is another contribution that watershed management can make to future land stewardship. However, these contributions will be difficult to achieve unless the adverse impacts caused by people on fragile riparian sites and the surrounding watersheds are reversed. A watershed management approach to land stewardship is a framework for more effective land-use polices.

5.1. FACTORS IMPACTING WATERSHED MANAGEMENT

Understanding key factors helps to establish a perspective about the contributions that watershed management can make to future land stewardship. Brooks and Eckman (2000) reviewed 30 watershed management projects in 20 countries spanning 1967 to 1999 to determine the key factors that contributed to project success or the barriers confronted in the transition from status quo to the sustainable use of land, water, and other natural resources (Table 5.1). The review ranged from small-scale, low-budget projects implemented by nongovernmental organizations (NGOs) to projects undertaken by international agencies with large regional focus and budgets greater than U.S. $250 million. The outcomes of early watershed management projects ranged from significant technical benefits to failures that caused unwanted environmental and sociocultural consequences.

Key factors that determined the outcomes included:

- Planning aspects;
- Scale and topography;
- Management and administrative issues;
- Tenure issues; and
- Institutional and policy implications.

Brooks and Eckman (2000) summarized the key factors in the following overview context.

Table 5.1. Watershed management projects reviewed and countries involved.

Project Title	Countries Involved
Interregional Project for Participatory Upland Conservation and Development	Bolivia, Nepal, Tunisia
Watershed Program in Andhra Pradesh	India
Watershed Program in Orissa	India
Watershed Program in Madhya Pradesh	India
Integrated Rural Environment Program	Java, Indonesia
Peum Perhutani Project	Java, Indonesia
Watershed Management Through People's Participation and Income Generation	Java, Indonesia
Yallah's Valley Land Authority Programme	Jamaica
Farm Development Scheme	Jamaica
Integrated Rural Development Project	Jamaica
Hillside Agricultural Programme	Jamaica
Agroforestry Development in NE Jamaica	Jamaica
Agricultural Production and Support Systems for Achieving Food Security	Grenada
Maissade Integrated Watershed Management Project	Haiti
Pilot Project in Watershed Management in the Nahal Shikma	Israel
Mae Se Integrated Watershed and Forest Use Project	Thailand
Salto Grande Hydroelectric Project	Argentina and Uruguay
Abary Water Control Project	Guyana
Cauca River Regulation Project	Columbia
La Fortuna Hydroelectric Project	Panama
Pueblo-Viejo-Quixal Hydroelectric Project	Guatemala
Tavera-Bao-Lopez Multipurpose Hydro Project	Dominican Republic
Kandi Watershed and Area Development Project	Punjab, India
Integrated Watershed Development Project	Himachal Pradesh, India
Loess Plateau Watershed Rehabilitation Project	China
Integrated Rural Development Through Communes	Rawanda
Women's Development in Sustainable Watershed Management	Myanmar
Sustainable Agriculture Development and Environmental Rehabilitation in the Dry Zones	Myanmar
Watershed Management for Three Critical Areas	Myanmar
Konto River Watershed Project Phase III	Indonesia

Source: Adapted from Brooks and Eckman (2000).

5.1.1. Planning Aspects

Most of the watershed management projects reviewed by Brooks and Eckman (2000) emphasized top-down planning, with technically-oriented objectives such as erosion control, reforestation, or afforestation. People who did not reside in the region planned and implemented many regional projects. These individuals had limited responsibility for project implementation and limited accountability for long-term outcomes. Participatory-planning methods (Tillman, 1981; Brooks et al., 1997; Ingles et al., 1999) were sometimes emphasized.

The importance of local involvement in the planning process was crucial to ensure that sociocultural dimensions were considered in project design. Watershed management projects that were technically oriented focused on outputs, while projects that used participatory methods focused on outcomes. More sustainable outcomes were promoted through close collaboration with local stakeholders.

Only a few of the projects attempted to incorporate land use, tenure issues, or other sociocultural factors before project implementation (Brooks and Eckman, 2000). More commonly, sociocultural studies were conducted when problems had surfaced after the planning process was completed and the project was operational. Once operational, NGOs helped some large projects diagnose the local sociocultural problems and they assisted in project management. Watershed management projects that used top-down planning without local input, frequently had unwanted technical, environmental, and sociocultural consequences.

5.1.2. Scale and Topography

Project success is influenced by how scale and topography were interrelated (Brooks and Eckman, 2000). Less ambitious projects designed for small watersheds were more successful in satisfying their objectives than large, more complex projects on widespread watersheds. The positive impacts of small projects focused on their immediate relevancy to local people. For example, improvement in crop production or farm income, rather than the number of gully-plugs constructed or miles of roads improved. Large projects, often encompassing many watersheds, reported complex impacts that were difficult to translate into effects at the local level.

Watershed management projects that focused on mountainous areas (Nepal, India, and Myanmar) or islands (Granada, Jamaica, and Java) possessed unique characteristics considered necessary to evaluate their contributions to the well being of people. The hydrologic response of mountainous watersheds to planned interventions was direct and often severe to upland and downstream inhabitants (Brooks and Eckman, 2000). While these watersheds were often prone to excessive rainfall resulting in landslides and large debris flows (Brooks, 1998; Cheng et al., 2000; Sidle, 2000), a scarcity of land, water, and other natural resources also prevailed. The capacity of fragile uplands to support populations of poor people and their animals was generally limited. Affluent people typically lived in the lowland areas. Therefore, differences in the socioeconomic conditions of upland and lowland inhabitants were an issue to consider within the framework of watershed management projects.

Extreme meteorological events exacerbated watershed management problems in areas that were montane and island (Brooks and Eckman, 2000). Cumulative watershed effects were often severe within these situations, with local communities experiencing direct and immediate consequences of flooding, landslides, and debris flows (Cheng et al., 2000; Sidle, 2000; others). Mountainous and small-island ecosystems, with densely inhabited watersheds, were particularly challenging to watershed management. At times, the proximity of uplands to productive lowlands and estuaries showed the upstream-downstream relationships that must be considered in watershed management. The inherent scarcities of land, water, and other natural resources confronted by the dense populations of people and their animals on small island watersheds, fostered conflicts over land and natural resource use. The main problem centered on the availability of a relatively limited resource base to sustain watershed management and other natural resource interventions.

5.1.3. Management and Administration

Management and administration are interrelated with other factors affecting the success of the watershed management projects. Brooks and Eckman (2000) found that small projects experienced better coordination, communication, and local participation than large complex projects. The small, focused projects were more successful in achieving their objectives than large projects. Watershed projects with minimal management and administrative structures had more flexibility than larger projects and, consequently, had greater success in monitoring and evaluating project benefits. The large projects also lacked mechanisms for the equitable sharing of project-derived benefits—some did not attempt to monitor these benefits.

Wages or other cash payments, food-for-work, and other forms of incentives (Bortero, 1986; Gregersen et al., 1987) were components of many watershed management projects. Although these incentives were intended as direct benefits for the project's participants (Brooks and Eckman, 2000), it was questionable whether they contributed significantly to long-term participation by local people. Maintenance of soil conservation structures or tree planting on degraded watershed sites often stopped when the incentives ceased. Therefore, the planners had to consider carefully whether incentives would effectively motivate the local people to sustain the practices necessary to achieve the watershed management goals and objectives.

Monitoring and evaluation were essential tools that allowed planners, managers, and decision-makers to track the progress of watershed management projects and adjust project execution to achieve the goals and objectives (Brooks and Eckman, 2000). Often, monitoring and evaluation were underused and under appreciated. Consequently, the mandated evaluations of many watershed projects were incomplete or incomprehensive and managers were unable to determine the success of the watershed-project components. The involvement of local-resource users and other watershed inhabitants in the monitoring and evaluation efforts facilitated more effective project management.

5.1.4. Land and Resource Tenure

Land and resource tenure is important to the sustainability of watershed and other natural resource management projects (Cernea, 1981; Gregersen at al., 1987; Raintree, 1987; Brooks et al., 1997). However, tenure issues were often neglected in the planning process for the projects reviewed (Brooks and Eckman, 2000). Tenure arrangements must be recognized to achieve sustainable projects and, ultimately, sustainable development and use of watershed resources. For example, in some countries, the forests and trees belong to the central government. When confronted by this situation, planners of watershed projects that promote tree planting to control erosion must understand how this would affect the local people and how these people might respond. Other questions that should be addressed in planning watershed projects included, "Who has the water rights and how are conflicts resolved?" Usually, tenure arrangements were seen as barriers to achieving project objectives for the projects reviewed and, when encountered, institutional and policy changes were recommended (Brooks and Eckman, 2000).

5.1.5. Institutional Support and Policy Implications

Appropriate policy and institutional support are essential for watershed management projects to be effectively integrated into long-term developmental programs that have lasting impacts on people and their use of land, water, and other natural resources (Brooks et al., 1994; Quinn et al., 1995; Cortner and Moote, 1999; 2000). Brooks and Eckmand (2000) found that policy and institutional implications involved all aspects of land, water, and other natural resource use and the concerns of stakeholders at all levels. They observed a lack of institutional memory regarding the lessons learned from earlier watershed management projects and an over reliance on repeating past methods without adapting them to changing circumstances. Being familiar with the effectiveness of watershed management and land stewardship institutions for achieving project success was important for decision-makers. Brooks and Eckman (2000) found that in most of the watershed management projects reviewed, continuously updating and increasing the stakeholder's understanding about watershed management was necessary. Development of effective mechanisms for maintaining the continuity of watershed projects and programs was essential to enable the information gained from past projects to be available for future reference.

Institutional support was required for project outcomes to be implemented into sustainable watershed management programs (Brooks and Eckman, 2000). The following two related points reinforce the advantages of a watershed management approach to land stewardship.

- Although the institutions responsible for management are organized around politically determined boundaries, institutional arrangements are necessary to ensure that management of land, water, and other natural resources recognize watershed boundaries.
- Interdisciplinary approaches are needed to manage the soil and water resources within a watershed framework.

Responsible agencies usually had specified mandates for management of a particular natural resource component, such as forest and trees, irrigation water, or hydropower, and were staffed with professionals in a particular discipline, such as foresters or engineers. However, these agencies generally lacked the authority or ability to cope with many watershed-level issues effectively, and they were not structurally organized around watersheds. This led to duplications of effort or voids in responsibilities. Community-based watershed partnerships and other groups in the United States (Lant, 1999) and elsewhere in the world (Ewing, 1999; Porto et al., 1999) have recognized this problem. However, institutional arrangements are necessary to facilitate management of land, water, and other natural resources in concert with each another.

5.1.6. Strategies for Sustainable Watershed Management Projects

In the past, water-resource-, forestry-, and agricultural- development projects have been planned and implemented with little regard for watershed management and the upstream-downstream relationship of a watershed management approach (Gregersen et al., 1987; Brooks et al., 1992, 1997; Brooks and Eckman, 2000). However, to develop sustainable projects and programs, land, water, and other natural resources must be managed together. Fortunately, in many countries educated or trained watershed managers are facilitating this by assuming leadership positions.

Emergence of locally-led watershed partnerships, councils, and boards (see Chapter 3) also suggests that societies are recognizing that, while a watershed management approach to land stewardship is relevant, institutions responsible for watershed management are not necessarily fulfilling their role. Nevertheless, this indicates that institutions and policies that support a watershed management approach are emerging.

5.1.6.1. Planning

Brooks and Eckman (2000) noted that during project design, interdisciplinary approaches that integrated the technical and human dimensions of watershed management were essential. Watershed planning has historically relied on technical and engineering expertise, but it has been largely deficient in the sociocultural aspects of planning. This has resulted in less than optimal project outcomes and in diminished benefit flows after project completion. Socioeconomic research and participatory techniques should be incorporated into the early stages of watershed management project design and planning (Gregersen et al., 1987; Brooks et al., 1990; Brooks and Eckman, 2000). Bringing socioeconomic specialists and local participants into the planning process after problems arise can be too late and places undue responsibility on those unfamiliar with the project design and plan.

5.1.6.2. Scale and Topopgraphy

Project design and planning should consider scale and topography to cope with upstream-downstream relationships and interactions and the possible cumulative watershed effects of the project (Reid, 1993; Quinn et al., 1995; Brooks et al., 1997; Sidle, 2000). Small-scale projects, with clearly defined goals and objectives, are more likely to attain beneficial outcomes that can lead to long-term watershed management programs than large, complex projects that are difficult to manage and administer (Brooks and Eckman, 2000). Gaining local participation is easier when the potential benefits of a project to the watershed's inhabitants are known, which is more commonly the case with small-scale projects, and when the impacts of the project are evident and appreciated (Tillman, 1981).

5.1.6.3. Management and Administration

Before using wages or cash payment, food-for-work, or other incentives to induce people to become involved in a project, alternative means of gaining local participation should be considered. Negative externalities can result when a project relies on incentives. Furthermore, incentives and similar economic strategies might not effectively conform to the participatory planning process because of economic and cultural differences between the local people and personnel of the responsible institutions.

Another point made by Brooks and Eckman (2000) regarding the management and administrative aspects of watershed management projects concerns the nature of the monitoring[28] and evaluation[29] efforts, which are conducted for different purposes. The two efforts must be effectively integrated to achieve maximum benefits. Furthermore, watershed management projects reviewed indicated that environmental and socioeconomic monitoring and evaluation are important to the success of a watershed management project (Gregersen et al., 1987; Brooks et al., 1990; Brooks and Eckman, 2000; Thorud et al., 2000).

5.2. WATERSHED MANAGEMENT CONTRIBUTIONS

By knowing the outcomes of watershed projects that have succeeded and of those that have failed to meet their goals and satisfying their objectives, the contributions of a watershed management approach to land stewardship can be effectively evaluated. The contributions are many, as elaborated by the participants in the conference on "Land Stewardship in the 21st Century: The Contributions of Watershed Management" and throughout this book. These contributions include:

- Providing an improved and logical basis in planning for future land stewardship;
- Helping to secure dependable supplies of clean water; and
- Forming an effective framework for sustaining the flows of crucial natural and economic resources.

Other noteworthy contributions are:

- Ensuring the stability of watershed-riparian systems that are vital to the sustainability of the benefits obtained through effective watershed management;
- Offering a diversity of investment, employment, and income opportunities; and
- Forming a foundation for effective land-use policies.

The following discussion about watershed management contributions to land stewardship focuses on the United States however, the contributions often apply to worldwide situations.

5.2.1. An Improved and Logical Planning Basis

The basic concept behind a watershed management approach to land stewardship is that it is an improved and more logical way to plan than using political or administrative boundaries (Hamilton, 1985; Gregersen et al., 1987; Brooks et al., 1992; Lopes et al., 1993; Brooks and Ffolliott, 1995; Thorud et al., 2000) (Box 5.1). As worldwide competition for land, water, and other natural resources intensifies, efficient land stewardship planning will become increasingly important. Many factors—the concept itself, the evolution of laws and regulations about natural resource management, and monitoring for compliance with the laws and regulations—support the watershed concept as the planning basis for land stewardship.

---

[28] The process of obtaining information to provide a basis for adjusting or modifying a project that has already been implemented to identify procedures for project continuation and improve future monitoring efforts.

[29] The process of appraising the results of a decision through the information collected from monitoring activities.

Box 5.1 Watersheds as Planning Units: A Historical Case Study

An example of the validity of using watersheds as planning and management units is found the history of ancient Hawaii (Morgan, 1985, 1986). The Polynesians, who settled the islands, organized their lives on the basis of a ahupuaa—watersheds that extended from the highest mountain peaks to the coast, including the coral reefs below the watersheds' ocean outlets. The ahupuaa was a natural region, in which sufficient natural resources to sustain the community existed. The ahupuaa was also a political entity under the control of an alii (chief), who was an environmental manager and political ruler. Since the Hawaiians were dependent on the products of the land, they manage the ahupuaa prudently for forests, the moderate slopes for upland crops, and planted the lowlands with taro. They used the streams to irrigate the taro, without polluting the fish-rearing coral reefs. The alii recognized that wise land-use practices, which avoided erosion and pollution, meant wealth for the political entity under their control. The ahupuaa system worked well before pre-European contact. However, the system began to break down after 1778, when Captain Cook opened the islands to Western influence and introduced the Hawaiians to a trading economy that eventually supplanted their subsistence economy.

5.2.1.1. Concept and Process

People are becoming increasingly familiar with the concept of a watershed when planning for natural resource management. They are aware of the physiography by which the ridge-lines of a watershed are defined; they understand the downward and cumulative flows of streams, rivers, and groundwater resources; and they recognize the general relationship between how much precipitation falls on a watershed and the resulting high and low streamflow regimes (Hamilton, 1995; Brooks et al., 1992, 1994, 1997; Thorud et al., 2000). This understanding is the greatest where topographic relief is well defined, as in the world's mountainous regions. There is also economic logic in considering a watershed a basic unit in the planning process, since it internalizes many off-site effects involved in land and natural resources.

A logical starting point for the planning effort using a watershed basis is before a problem or opportunity has been identified (Brooks et al., 1990, 1997). Therefore, the planning process ideally begins with monitoring and problem identification. People are continuously collecting data and creating knowledge about ongoing changes in biological, physical, and socioeconomic environments on watershed lands. This information is useful in the initial stages of planning to decide whether a problem exists and, if so, what might be done to resolve it. Despite how a problem or set of problems is ultimately recognized, its articulation is the first step in the planning process. The general sequence of subsequent steps in the circular planning process is shown in Figure 5.1.

5.2.1.2. Laws and Regulations

As in many other countries, the laws and regulations in the United States encourage planning on a watershed basis. The Endangered Species Act of 1973 requires that suitable habitat conditions be available to ensure that endangered and threatened plant and animal

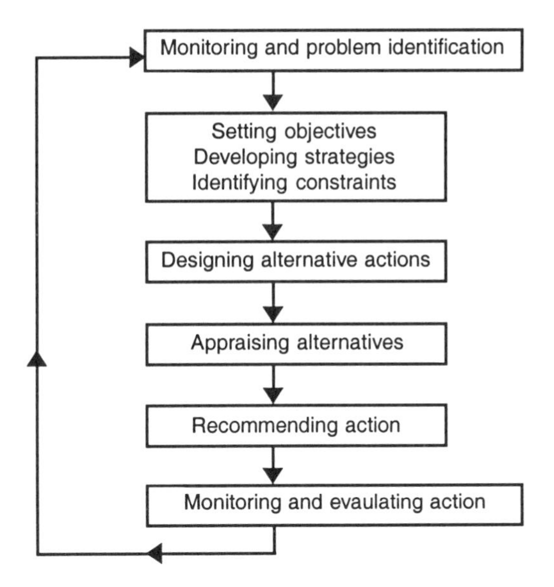

Figure 5.1. The general sequence of steps using a watershed management approach to land management planning (from Ffolliott et al., 1999. Copyright 1999 by the Board of Trustees of the University of Illinois).

species survive. Critical habitats on watershed lands play an important role in meeting the specifications of this law (Cortner and Moote, 1999; Beschta, 2000; Thorud et al., 2000). Protection and mitigation efforts for endangered and threatened species require that land users, including those engaged in forestry, range management, agriculture, utilities, and urban development activities, consider their own and the combined impacts of others on a watershed. The planning process must be coordinated across all land uses and ownerships. Otherwise, the efforts of one landowner or resource user to recognize the on- and off-site effects and to mitigate the impacts to endangered and threatened species will be compromised by the activities of others. Even disturbances on small areas of a watershed can adversely affect the downstream water supply and quality. Therefore, desired outcomes will not be attained if the major factors influencing species survival, water supply and quality, and other natural resources within the watershed are not adequately addressed in the planning process.

The Clean Water Act of 1972 is another federal law in the United States with direct implications for planning on a watershed basis (Neary, 2000; Thorud et al., 2000). The primary goal of this act is implementing watershed management practices to maintain or improve the physical, chemical, and biological integrity of the country's water. For example, this law could regulate the total maximum daily load (TMDL) of sediments entering a stream system (Brown et al., 1993; Gelt, 2000; Tracy et al., 2000). Because the total sediment load at any point in the stream depends on everything that influences the erosion and sediment dynamics above that point—particularly upstream land-use practices—the individual and collective influences of the upstream-land uses must be addressed. However, TMDL regulations only mitigate a problem after it has occurred. Therefore, regulatory or voluntary best management practices (BMPs) are necessary to meet these regulations.

State laws and regulations that encourage the watershed as a planning basis are emerging throughout the United States. These actions are frequently related to the need to protect an endangered or threatened fish species or to meet the needs of stakeholders, such as farmers and residents of municipalities, who rely on out-of-stream water uses (Beschta, 2000; Thorud, 2000). Provisions of these laws and regulations are often voluntary and can require representation from state agencies, local governments, representatives of the area's economic interests, and the public. The main goal of the laws and regulations is collaboratively to develop integrated management plans for land, water, and other natural resources on a watershed basis.

5.2.1.3. Monitoring

Monitoring for watershed management compliance with federal and state laws is another reason a watershed is a logical basis for planning. Streamflow and water quality characteristics that are systematically monitored at gaging stations on rivers and their tributaries provide measures of compliance with the laws and regulations. If obtained over a long time, the characteristics also indicate a watershed's response to changes in policy or management practices (Brooks et al., 1997; Sidle, 2000) and the cumulative effects of the land-use practices on the watershed (Reid, 1993; Brooks et al., 1997). Monitoring techniques are being developed to track the movement of particles, pathogens, fish, and other elements of a watershed that indicate the environmental health of the watershed. Additionally, monitoring may pinpoint the sources of pollution. However, it is more important that watershed monitoring help people to understand the ambient health of their environment, and potential impacts of their actions.

5.2.1.4. Status of Planning on a Watershed Basis

Planning to assess and manage the cumulative effects of all resource uses and land-use practices on watershed lands requires the involvement of all landowners and resource users within the watersheds (Reid, 1993; Brooks et al., 1990; 1997; Sidle, 2000; Thorud et al., 2000). Legal restrictions on land, water, and other natural resource management efforts assume that landowners will not voluntarily participate to achieve watershed management or land stewardship objectives. Laws, regulations, and public sentiment compel landowners to meet society's expectations for the watershed—this was true for air-quality management of critical airsheds in the 1970s. Unfortunately, impediments to successful planning on a watershed basis still exist. Among the more significant of these barriers (Thorud et al., 2000) are:

- Landowners with different goals and objectives that conflict with the public's watershed management goals;
- Years of regulatory behavior that has not rewarded collaborative planning efforts;
- Confusing federal and state agency responsibilities;
- Incoherent and disparate data collections, or distrust of data;
- Multiple political jurisdictions of watershed lands;
- Undirected or insufficient funding, or funding cycles that are too short to address the problems;
- Invalid simulation models; and
- Lack of understanding of hydrologic processes and watershed management.

Despite these difficulties, we anticipate that significant progress in using a watershed as a planning basis to achieve improved land stewardship will occur in the future. Needed changes in policy and practice will not be revolutionary but evolutionary and, hopefully, increasingly progressive and effective.

5.2.2. Securing Clean Water

Securing clean water will always be a primary goal of watershed management practices, projects, and programs. Responsible management agencies have traditionally taken a regulatory approach to achieving this important goal (Clarke and McCool, 1996; Cortner and Moote, 1999; Somerville and DeSimone, 2000). However, greater emphasis is being placed on local-level decisions and actions to ensure clean water on a sustainable basis. Conservation districts, locally-led advocacy organizations (see Chapter 3), and other local groups have assumed increased roles and responsibilities for ensuring the delivery of high-quality water supplies through watershed management. No matter how it is accomplished, a major goal of a watershed management approach to land stewardship in the 21st century is to secure flows of high-quality water to all stakeholders in the river basin.

5.2.2.1. Background Legislation

Periodic amendments to federal water-quality legislation in the United States provide specific guidelines for watershed management and water-quality protection. Sometimes, the amendments have extended watershed and water-quality protection to state and private lands (Brown et al., 1993; Cortner and Moote, 1999; Beschta, 2000; Neary, 2000). The 1972 amendments to the Federal Water Pollution Control Act—commonly called the Clean Water Act—established a significant federal presence in controlling water quality in the United States. The Clean Water Act is the primary law that addresses the protection of all lakes, rivers and streams, aquifers, and coastal areas in the United States.

The main objective of the Clean Water Act is to provide the mechanisms for maintaining or restoring the integrity of the country's waters. This objective translated into the fundamental goals of achieving water-quality levels that promoted fishing and swimming by 1983 and eliminated the discharge of all pollutants into these waters by 1985 (Brown et al., 1993; Beschta, 2000). The Clean Water Act provides a comprehensive framework of standards, technical tools, and financial assistance to address the causes of water pollution and poor water quality, including municipal and industrial wastewater discharges, polluted runoff from rural and urban areas, and habitat destruction. For example, the act:

- Requires major industries to meet performance standards to ensure pollution control;
- Charges states with setting appropriate, specific water-quality criteria and developing pollution control programs to meet them;
- Provides financial assistance to states and local communities to help them meet their clean-water requirements (Box 5.2); and
- Protects valuable wetlands and other aquatic habitats through a permitting process that ensures that development and other activities are conducted in an environmentally sound manner.

Box 5.2 State Involvement to Meet Clear-Water Requirements: A Report

The conclusion of a report titled, *Pollution Paralysis II: Code Red for Watersheds*, is that four-fifths of the states in the United States do not meet Clean Water Act standards in their attempt to protect people from contaminated rain and runoff. This publication is a sequel to a 1997 National Wildlife Federation report. This report analyzed each state's use of a watershed restoration approach, as mandated by the Clean Water Act, to control the diffuse source of water pollution (International Wildlife, 2000). Contaminated rain and runoff are polluting more than 480,000 kilometers of rivers, streams, and shorelines and nearly 20,000 square-kilometers of lakes. The Clean Water Act requires that states place limits on all sources of pollution that enter impaired waterways and take steps to ensure that the limits are not exceeded. According to the *Pollution Paralysis II: Code Red for Watersheds* report, states have focused their attention on the pollutants discharged from pipes, rather than on contaminated rain or polluted runoff.

In 1972, point-source emissions were the major emphasis of the Clean Water Act. However, Section 208 specifically addressed nonpoint source pollution and designated silvicultural treatments, such as timber harvesting, improvement thinning, road construction, and livestock grazing, as significant sources of nonpoint source pollution (Brown et al., 1993; Brooks et al., 1997; Beschta, 2000; Gelt, 2000). Important to forestry was Section 404, which focused on water pollution associated with the deposit of dredged and fill materials. Although the U.S. Environmental Protection Agency—the agency responsible for enforcing Clean Water Act provisions—initially placed emphasis on problems of point-source pollution, they also suspected that water-quality degradation from nonpoint source pollution would be a major barrier to achieving the Act's water-quality goals (U.S. Environmental Protection Agency, 1974; Brown et al., 1993). Among other provisions, the amended Clean Water Act of 1977 authorized a program of grants to help rural landowners implement BMPs to control nonpoint source pollution (see Chapter 4). BMPs are usually the most practical approach to controlling nonpoint source pollution, complying with water-quality standards, and securing clean-water flows.

5.2.2.2. Best Management Practices (BMPs)

Many BMPs for silvicultural treatments, livestock-grazing practices, road-related soil disturbances, and agricultural activities, which mitigate erosion-sedimentation processes, are known (Moore et al., 1979; Lynch et al., 1985; LaFayette et al., 1992; Chaney et al., 1990; Verry et al., 2000). However, BMPs for other pollutants are unknown or incomplete—research relating land use to water quality is needed. Monitoring efforts reveal that BMP use is increasing and that implementation usually maintains water quality (Brown et al., 1993). Information about the overall cost effectiveness of a BMPs program is sometimes lacking.

The intent of most BMPs is consistent with improved watershed management practices because little concern exists about nonpoint source pollution from watersheds in acceptable condition (Box 5.3). However, specifying the BMPs to achieve an unmet water-quality standard requires understanding the relationship between the disturbances caused by land-use practices and downstream water quality (Brooks et al., 1997). This complexity exists because distinguishing among the individual upstream causes of nonpoint source pollution

Box 5.3 Best Management Practices for Eastern Riparian Ecosystems

Best Management Practices (BMPs) specific to riparian ecosystems in the Eastern United States include limiting disturbance and excluding pollutants and establishing guidelines for waterway crossings. BMPs are designed to mitigate or prevent impacts due to changes in streamflow, sediment movement, water temperature shifts that affect aquatic habitats, and input of chemicals, organic debris, and solid waste (Verry et al., 2000a). However, before any natural resource management begins, the landowner or manager must specify their management objectives. Once these objectives are known, the specific BMPs needed to accomplish the objectives can be planned.

Planning, often the most important BMP, is an opportunity to identify site-specific needs, recognize landscape-level concerns, potential problems, and conflicts, and select activities that prevent adverse impacts, modify the intensity of these impacts, or improve poor conditions. It is important to consider where and how much disturbance is acceptable, the extent of the impact area, the location and type of mitigating practices, and the best season for the activity. Specific BMPs can exclude or control activities that disturb the soil, such as roads, skid trails and log landings, boat landings and other concentrated recreational sites, and fire and fire lines. Use of tractors or other heavy machinery in implementing management activities also might be excluded. The number, size, or location of trees that may be cut to meet the management objectives is often limited to reserving trees for shade strips, protecting aesthetic views, or furnishing coarse woody debris to the stream system.

are difficult and separating naturally occurring water-quality degradation from management caused is demanding (Brown et al., 1993). A watershed management approach to land stewardship helps overcome these conflicts.

5.2.2.3. Clean Water Initiative

In early 1998, the Office of the President announced a Clean Water Initiative to meet the clean-water goals of the United States (U.S. Environmental Protection Agency, 1999; Towns, 2000). A key part of the initiative is planning and implementing locally-led watershed protection efforts in high-priority areas. Focusing these efforts on the watershed establishes the priorities for protecting and restoring the water quality. To revitalize the commitment to share water resources, the Forest Service, Bureau of Land Management, Bureau of Indian Affairs, and other federal agencies that have water and watershed management responsibilities, were directed to prepare a comprehensive Clean Water Action Plan to accelerate the progress in improving water quality. The action plan is designed to provide clean water, public-health protection, and sustainable water resources for the communities. Federal agencies involved in the initiative are working closely with state and local partners to implement the Clean Water Action Plan.

5.2.2.4. Unified Federal Policy

Responsible federal agencies are also developing a Unified Federal Policy for ensuring a watershed approach to management of federal lands and natural resources (U.S. Environmental Protection Agency, 1999; Towns, 2000). This proposed policy is designed to:

- Assess the function and conditions of watersheds;
- Incorporate watershed management goals into the planning process;
- Enhance pollution prevention and control;
- Monitor and restore the watershed;
- Recognize water bodies of exceptional value; and
- Expand planning and managing collaboration.

The draft policy statement outlines federal commitments to achieve consistent watershed assessments, improve watershed management practices, comply with water-quality requirements, and enhance collaboration with the states and stakeholders. It also provides a time line to meet these commitments. When formalized, this policy will be a valuable instrument to help ensure the security of clean water at the watershed scale.

5.2.3. Sustaining Flows of Natural and Economic Resources

Watersheds are the source of many natural and economic resources that benefit society. These resources are often subject to joint production processes[30] (Gregersen et al., 1987; Brooks et al., 1992; 1997; Ffolliott and Brooks, 1998; de Steiguer, 2000; Towns, 2000). Because of the processes, tradeoffs of crucial resources are often necessary. The optimal amount of goods and services derived from these shared resources should be resources that maximize the present value of net benefits to society. The best way to achieve this is through integrated resource planning on a watershed basis.

5.2.3.1. Natural Resources

A variety of natural resources, which sustain economic development and benefit human health, are found on watershed lands (Gregersen et al., 1987; Brooks et al., 1990; 1997; World Resources Institute, 1996; de Steiguer, 2000; Eckman et al., 2000; Towns, 2000). Rivers emanating from upstream watersheds are often transportation corridors carrying a variety of commodities to the marketplace. Most of the world's supply of clean water originates on upland watersheds, some of which are far from the end users. Water quality, another aspect of a watershed that is important to people, depends on natural and human factors—human impacts present a significant threat to human welfare (National Research Council, 1999). Erosion, sediment, and flood control and hydropower generation are other watershed-based activities that benefit society. Other natural resources on watershed lands that benefit society include wood, forage, livestock, and agricultural crop production and recreation (Box 5.4).

5.2.3.2. Economic Resources

While describing shared natural resources on watersheds in physical terms is useful, the economic view also provides an informative perspective (Gregersen et al., 1987; de Steiguer, 2000). The economist's view of the value of natural resources is fundamental to develop a

[30] The production or use of one resource is related to the production or use of one or more of other resources.

Box 5.4 Natural Resources on Watershed Lands

Forests and woodlands cover many watersheds worldwide and, therefore, are the timber source for processing a variety of wood products (National Research Council, 1999; de Steiguer, 2000). The relationship of timber harvesting to streamflow regimes has long been an issue of interest to watershed managers, foresters, and other land managers (see Chapter 3). Agricultural crop production and livestock grazing are also watershed activities important to many people—these activities are often combined with tree growing in agroforestry practices to benefit local inhabitants (MacDicken and Vergara, 1990; Nair, 1993; Gordon and Newman, 1997; Garrett et al., 2000). Hunting and fishing and non-consumptive values (e.g., viewing and photography) are also found on watershed lands. Watersheds are sites of many opportunities for outdoor recreation that provide public benefit. Important local natural resources also come from watershed lands. Examples of these include the cranberry harvests and dairy production on the Willapa Bay Watershed in Oregon and the crab and oyster production on the Chesapeake Bay (National Research Council, 1999; de Steiguer, 2000; Verry et al., 2000b). Watershed lands are the source of many unique natural resources that benefit people in the United States and worldwide.

strategy for sustaining resource flows. Use values[31] are separated into market values[32] and nonmarket values[33], while economists use non-use values[34] to describe natural resources on watershed lands (Box 5.5). These values should be quantified to the extent possible to assess the benefits and costs of watershed management in sustaining the flow of crucial resources.

The array of economic resources considered by economists is not always included in the resources on a watershed. Nevertheless, they are resources with social values that must be considered when managing the flows of watershed resources for sustainability (de Steiguer, 2000). Many planners, managers, and decision-makers currently recognize the relevance of nonmarket and non-use resources to natural resource decisions. Various methods for considering the monetary value of nonmarket and non-use resources have been developed in recent years to help facilitate this decision-making process (Gregersen et al., 1987; Bromley, 1995).

5.2.3.3. Sustaining Flows

Traditionally, a watershed is planned and managed to obtain equal periodic flows of resources over time (de Steiguer, 2000). Recently, the idea of managing the flows of economic resources on a watershed has replaced the philosophy of an even flow of natural resources. The economic goal of this multiple-use problem is to select a sequence of watershed management actions that maximize the present value of the net benefits from the combined

[31] The values derived by the direct consumption or use of a resource, rather than only contemplating or enjoying the resource passively.

[32] Values of resources that are traded in the marketplace.

[33] The value of resources that are intangible, such as knowing that the resource will exists for future generations; no payment is made to the producer for the resource and the value is not reflected by a transaction in the marketplace.

[34] Resources with values derived from the possibility of their future use, not from their current use, and not necessarily by the person currently conveying a value to the resource.

Box 5.5 Value of Natural Resources on Watershed Lands

Use values include those for water, timber, and the other natural resources available on a watershed. Production costs of the seller and willingness to pay by the consumer establish the market value, therefore, the price of the good or service. Hydro-power, timber, and some forms of recreation, such as paying for camping privileges, are examples of market commodities with use value. These resources are produced and consumed by users who pay a price that approximates their value.

Nonmarket values are also obtained from the direct consumption or use of the resource, however, as in flood control, there is no cash market for the good. Other examples of nonmarket values include wildlife observations, clear flowing water in streams, and views from scenic forest visas. Although intangible, these economic-resource values exist and are meaningful to society. The goods and services obtained from nonmarket values are public goods—if a producer has provided them, they have the technical characteristics of being indivisible and accessible to all (Gregersen et al., 1987; Tietenberg, 1988; de Steiguer, 2000).

Non-use values, the second categorization of economic values, include option values that reflect the willingness to pay to preserve the option of being able to use an amenity in the future and existence values that indicate a willingness to pay to know that a resource will exist in the future. Bequest values measure the willingness to pay to maintain a resource for use by future generations and stewardship values are those placed on managing a resource to maintain its environment health for use by all organisms.

flows of water, timber, and other resource values over time (Gregersen et al., 1987; Krutilla and Bowes, 1989; de Steiguer, 2000). The idea of maximizing the economic benefits of the resources on a watershed is evident in the United States in federal legislation such as the Forest and Rangelands Renewable Resources Planning Act of 1974 and the National Forest Management Act of 1976. Legislation mandates resource managers to:

- Maximize the net benefits from the multiple-use, ecosystem-based, and sustained-yield management of watershed lands;
- Consider the relative values of all resources; and
- Preserve the integrity of these lands (Cortner and Moote, 1999; de Steiguer, 2000).

Watershed management prescriptions to attain the goals are complex, detailed, and holistic in their implementation, and they require large amounts of data and analyses to prepare. Fortunately, emerging tools and technologies available to planners, managers, and decision-makers (see Chapter 3) help make these goals attainable.

Sustaining economically efficient resource flow from watershed lands is achieved through integrated resource planning efforts structured around a watershed approach to land stewardship. Analytical techniques, methods, and procedures are available to help in these planning efforts (Gregersen et al., 1987; Gregersen and Conreras, 1992; Loomis, 1993; Brooks et al., 1997; de Steiguer, 2000). However, comprehensive plans evolving from these efforts to produce the greatest value of goods and services for society over a specified period represent an ideal toward which watershed managers strive. Watershed managers have made progress implementing practices, projects, and programs to meet this ideal in recent years and, in doing so, are sustaining the flows of crucial natural resources from watershed lands.

5.2.4. Ensuring the Health and Stability of Watershed-Riparian Systems

Sustaining the health and stability of watershed-riparian systems is important to ensure the sustainability of the benefits obtained through effective watershed management. However, improper land use and management practices can threaten this health and stability. Human-induced impacts caused by ill-planned land-use activities on the surrounding watersheds can have a greater potential for introducing changes to the composition, structure, and function of riparian ecosystems than human-caused disturbances within riparian ecosystems (Federal Interagency Stream Restoration Working Group, 1998; Clary et al., 2000). While much has been written about the unique values found within riparian ecosystems, only recently have publications and technical reports described how land-use activities on the surrounding watersheds can affect the stability of these systems (DeBano and Schmidt, 1989; 1990; DeBano et al., 1996; Baker et al., 1998; Verry et al., 2000b; Garrett et al., 2000).

Intermittently low- and high- streamflow regimes coupled with the discontinuous storage and movement of sediments through the channel is a hydrological, complex phenomenon. Therefore, interpretation is difficult when assessing the responses of riparian ecosystems to upslope watershed management practices (Heede et al., 1988; DeBano et al., 1996; Baker et al., 1999; Whiting et al., 1999). Nevertheless, we know that implementing watershed management practices that sustain baseline streamflows without excessive fluctuations, minimize soil losses on the surrounding watersheds, and minimize the flow of sediments through the riparian ecosystems, can stabilize watershed-riparian systems.

5.2.4.1. System Dynamics

Knowing how degradation of riparian ecosystems occurs provides insight about how to prevent it through proper watershed management practices or how the stability of the systems can be restored if degradation has occurred. A common scenario leading to a shift in the balance between watershed condition and riparian health and stability (see Chapter 4), followed by degradation of a riparian ecosystem, often begins with excessive timber harvesting, livestock or wildlife overgrazing, or the occurrence of wildland fire on the surrounding watersheds (DeBano et al., 1996; Baker et al., 1998; Clary et al., 1996, 2000; Verry et al., 2000a). A loss of the protective vegetative cover and compaction of the soil surface throughout the watershed follows. Infiltration of net rainfall is reduced and overland flows are increased when plant removal and soil compaction are severe. Increased overland flows on the hillslopes deliver more water to the channel, increasing the high-flow regimes that can exceed the channel's capacity and cause channel enlargement and downcutting (Heede, 1980; DeBano and Schmidt, 1989; Frasier et al., 1995; Krueper, 1996). These activities expand the drainage networks that maintain the undesirable flash-runoff events and increase sediment delivery to the channel system.

When roads and trails are developed on the surrounding watershed, overland flows are further concentrated and cause additional water delivery to the stream channel (DeBano et al., 1996; Baker et al., 1998). A channel that becomes incised can intercept and drain the local water table, which might be close to the surface and supporting a healthy riparian ecosystem. When a water table drains and lowers, the result is dewatering[35], severe alteration,

[35] When a water table lowers to the point where a stream or river ceases to function as a perennial waterway.

and eventually ecosystem destruction and site productivity reduction (Baker et al., 1999; Clary et al., 2000; Verry et al., 2000a). On lower-elevation mainstreams, wood cutting, agricultural development, and urbanization further complicate the problems caused by upstream land-use practices and, consequently, are responsible for further riparian site destruction.

5.2.4.2. Management Strategies

To remedy the above scenario, implementing watershed management practices that reverse the processes of degradation is necessary. Excessive tree cutting, overgrazing, and catastrophic wildland fire must be reduced or excluded. Mitigation efforts must be planned, while remembering that upland watersheds and riparian ecosystems are physically, biologically, economically, and socially related (Baker et al., 1999; Clary et al., 2000; Ilhardt et al., 2000, Verry et al., 2000a). Effective restoration treatments that return degraded watershed-riparian systems to health, stability, and higher productivity can be efficiently planned and implemented within the framework of a watershed management approach to land stewardship. However, restoration of riparian ecosystems often requires special managerial considerations (Box 5.6). Knowledge of the fluvial, geomorphological, and ecological functions, processes, and dynamics of the watershed-riparian connection is also needed to sustain the health and stability of the systems.

5.2.5. Investment, Employment, and Income Opportunities

The sustainable flow of benefits from natural resources on watershed lands provides opportunities for on- and off-site investments. Because of these investments, employment opportunities and income generation are possible (Gregersen and McGaughey, 1985; Ffolliott and Brooks, 1986; Arnold and Flaconer, 1987; Quinn et al., 1995). Public investments from the budgets of the responsible governmental agencies contribute significantly to natural resource management investment in the United States and other countries. The magnitude of the public investment is large because the high-elevation watershed lands of the world are

Box 5.6 Riparian Restoration Treatments: Some Examples

Restoration of riparian ecosystems often requires elimination of tree cutting and a reduction of grazing by domestic animals and wildlife for a few years. This prescription would allow the riparian plants to regain their vigor and restore their ability to detain high streamflow regimes, reduce the energy of flowing water and consequent bank erosion, and trap sediments and nutrients in the water (Medina, 1996; Baker et al., 1998; Verry et al., 2000a). Often, constructing fences around sensitive, degraded sites to exclude livestock and promote vegetation recovery is necessary (Clary et al., 2000; Long, 2000). Installing larger culverts at road crossings to eliminate blockages by debris during runoff events might also be necessary. Riffle bars are occasionally reestablished in the channel to slow water velocities, reduce or end channel downcutting, and provide fish spawning habitats (Minckley and Rinne, 1985; Baker et al., 1999). Considering the connectivity of riparian ecosystems along stream corridors when implementing these and other restoration treatments is important. The effectiveness of these treatments can be compromised when large gaps of channels remain unprotected by a riparian buffer (Schultz et al., 2000).

mostly in the public domain and, therefore, provide a diversity of commodities and amenities to society. Private investments are also made in watershed management, although the perceived high risks and low returns often limit the levels of these investments.

5.2.5.1. Investment Opportunities

The multiple-use nature of a watershed management approach to land stewardship offers a variety of investment opportunities in timber, livestock, and agricultural crop production; water-delivery infrastructures; and recreational facilities. On-site marketable investment opportunities occur when planned production increases in wood, livestock, agricultural crops; when other commodities are sustained on watershed lands; and when soil and water resource conservation considers current and future production of these marketable commodities (Ffolliott and Brooks, 1986; Gregersen et al., 1987; Quinn et al., 1995). A nonmarketable on-site benefit of watershed management is an environmental enhancement, such as sustaining productive soil bodies, a sufficient amount of clean stream water, or desirable landscapes. Nonmarketable on-site benefits are gaining value as rural and urban developments continue to encroach on fragile sites within watersheds and upper river basins.

Investment opportunities also occur when protecting earlier off-site ventures that depend on a sustainable flow of high-quality water or other natural resources from upland watersheds (Ffolliott and Brooks, 1986; Gregersen et al, 1987; Brooks et al., 1992; Quinn et al., 1995). For example, investments made to stabilize soil on watersheds that are upstream of reservoirs often protect large downstream investments made in the:

- Construction and management of reservoirs that supply water for irrigated and often large-scale agricultural production;
- Dependable and sufficient supplies of drinking water for people and their livestock; and
- Generation of hydropower for industries and municipalities.

The reduction in sedimentation rates that protect or extend the economic life of the reservoirs and the investment made in downstream ventures often justifies investments in soil stabilization (Box 5.7). Too often, soil stabilization and other watershed management practices designed to sustain the flow of high-quality water to a downstream reservoir are not adequately considered until after the reservoir is constructed. At this point, it is usually too late to protect the investment made in the reservoir for its projected economic life.

5.2.5.2. Employment and Income Opportunities

A relationship often exists between natural resource management on watersheds and employment and income opportunities (Arnold and Falconer, 1987; Gregersen, 1988; Gregersen et al., 1989)—employment and income opportunities are often generated through watershed management programs. Watershed management programs provide employment and income opportunities—harvesting, collecting, or gathering and processing wood, livestock, wildlife, and agricultural crops, and operating recreational facilities and eco-tourism concessions—to stakeholders in the public and private sectors. The employment and income generated can be important to the rural poor living on the watershed lands. Watershed management activities can also stimulate employment opportunities in transportation and maintenance services. Income obtained in the public sector is often reinvested in:

Box 5.7 Protection of Downstream Investments with Watershed Management: A Case Study

An economic analysis of a watershed-management program that reduced the sedimentation of a reservoir on the Lower Agno River the Philippines indicated that the program could be justified if properly implemented. Based on estimates of sedimentation rates following reservoir construction, Briones (1985) found that the reservoir's predicted 50-year lifespan was too optimistic. Sediment would accumulate in the reservoir to reduce its operational effectiveness earlier than expected, causing significant economic losses in irrigation, flood control, and hydropower capacity. However, implementing vegetative and structural measures to control the soil-erosion rate could avoid these losses. If implemented, the watershed-management program would extend the reservoir's life and have a favorable benefit-cost ratio of between 2.73 and 4.14. Therefore, the reservoir could be justified from the perspective of downstream benefits alone. The economic benefits of the program would be even greater if the decision-makers considered the on-site benefits of sustaining or increasing upland productivity in the river basin.

- Administrative infrastructures;
- Monitoring and continuing evaluation of operational natural resource management programs on watershed lands; and
- Effective planning, implementation, and administration of future watershed management practices, projects, and programs to meet the public's increasing and often changing needs.

Income generated in the private sector is largely personal income for the individuals or groups of individuals involved in the efforts that are necessary to:

- Sustain the watershed management activities;
- Meet other contingencies or emergencies; and
- Satisfy their personal needs and desires.

A watershed management approach to land stewardship is a philosophy of natural resource management. It also represents a practical way to achieve watershed management benefits, while diversifying and often increasing the investment, employment, and income opportunities of the stakeholders living on and off the watershed lands (Ffolliott and Brooks, 1986; Brooks et al., 1992, 1997). By recognizing the multiple benefits obtained from watershed management and incorporating these benefits into comprehensive management of land, water, and other natural resources, efficient use of the natural resources on watershed lands is possible. Through a watershed management approach, attempts are made to optimize the on-site benefits of forestry, livestock, and agricultural crop production, wildlife harvesting, recreation, and tourism. Additionally, the off-site benefits, such as protecting the investments made in downstream enterprises and other ventures that provide a wide array of economic and financial benefits to people in many sectors, are increased.

5.2.6. More Effective Land-Use Policies

A watershed management approach to land stewardship provides a framework for effective land-use policies. The massive and often spontaneous migrations of people throughout the world and the planned resettlement of people to upland watershed lands have caused the exploitation of natural resources and the degradation of land and water (Gregersen et al., 1987; Brooks et al., 1994, 1997; Quinn et al., 1995). Resettlement often threatens the current and future livelihood and well being of millions of inhabitants upland and downstream on these watersheds. Policymakers face many challenges concerning these problems and issues. As mentioned, social and political institutions rarely coincide with watershed boundaries, which govern most of the biophysical processes that control the productivity and sustainability of land use. Furthermore, on watershed lands a diverse set of stakeholders representing a variety of downstream interests and political issues affect natural resource use on watershed lands.

5.2.6.1. Policy Development

While a watershed management approach provides a framework for policy formation to achieve a productive and sustainable use of natural resources on watershed lands (Box 5.8), policymakers should consider the following key factors when developing policy (Brooks et al., 1992, 1994; Quinn et al., 1995; Eckman et al., 2000).

- Identify all stakeholders, and understand their perceptions and attitudes about natural resource use and watershed management.
- Improve clarification and coordination among the agencies responsible for planning and implementing watershed management concerning the administrative responsibility and legal jurisdiction over and land-use activities on the watershed lands.
- Encourage participation of watershed inhabitants, often the most capable natural resource managers on these lands, through land-use policies.
- Develop policies to equitably distribute benefits to those who employ environmentally sound land-use practices, and correct those responsible for the degradation of natural resources.
- Improve assessment of the impacts of the current policy to design an effective future policy.

Ineffective land-use policies can lead to land, water, and other natural resource degradation (Brooks et al., 1992, 1994; Quinn et al., 1995; Eckman et al., 2000). The degradation can be contained, while the productivity of the uplands can be environmentally sound and the environmental effects of land, water, and other natural resources decisions can be comprehensively considered. This can occur by recognizing and reconciling the potential conflicts among differing policies, and their impacts on natural resources. Once this has been

Box 5.8 A Policy and Its Components

A policy is the establishment, implementation, and, if necessary, enforcement of the appropriate institutional arrangements to guide a land-use choice, which was made by the involved stakeholders (Bromley, 1992). An effective policy specifies how to manage the land-use practices selected for implementation andconsists of three related and essential components—intentions, rules, and compliance. Intentions reveal what the stakeholders hope to accomplish with the institutional arrangements. Rules are the institutional arrangements that constrain some people and liberate others. Compliance is the essential component that converts promises and proclamations into meaningful results.

accomplished, effective land-use policies can be established using a watershed management approach. Successfully implementing these policies is considered in Chapter 6.

5.2.6.2. *The Challenge*

Developing appropriate land-use policies is a challenge. Appropriate policies might exist, but they are inefficiently or ineffectively implemented (Brooks et al., 1994; Quinn et al., 1995; Eckman et al., 2000). It is important to first assess the current situation and then identify any weaknesses in the existing policies. Options available for overcoming the identified weaknesses will vary for each particular case. Additional education or training is needed for those who plan and implement practices, projects, and programs of land, water, and other natural resource use but do not understand watershed management solutions. Education or training may be needed by farmers, natural resource planners, managers, and decision-makers.

Building on local experiences is also important, recognizing that a range of situations exists. For land-use policies to achieve widespread land-use improvements and sustainability effectively, the social, economic, and political realities of the situation must coincide with the technical ability. Potential solutions will include many different technical and institutional measures. A watershed management approach to land stewardship provides a comprehensive and practical biophysical, socioeconomic, and institutional framework for policy development and for solving natural resource conservation and sustainability problems.

5.3. SUMMARY

The contributions of watershed management to land stewardship are extensive. A watershed provides an improved and logical basis for planning, while internalizing many downstream impacts of upland watershed practices, projects, and programs. Watershed management represents a functional landscape encompassing technical interrelationships and interdependencies necessary for land, water, and other natural resource management. Helping to ensure the delivery of high-quality water supplies to people is another major contribution of watershed management. Securing clean water will become even more critical on a global-scale in the coming century because of the world's increasing human population.

A watershed management approach to land stewardship is an effective framework for sustaining the flows of crucial natural and economic resources to people. The optimal amount of goods and services derived from the shared resources on watershed lands should be the flow of resources that maximizes the present value of the net benefits of these resources to society. Often the best way to achieve this maximum is through integrated and participatory planning on a watershed basis. Another important contribution of watershed management to land stewardship is maintaining the health and stability of watershed-riparian systems—vital to sustaining benefits obtained through effective watershed management. Watershed management also enhances opportunities for investment, employment, and income of people living on and off the watershed lands. A diversity of these opportunities is presented through the multiple-use nature of the watershed management approach. This approach can also be a foundation for the development of effective land-use policies to manage natural resources for improved land stewardship.

REFERENCES

Arnold, J. E. M., Falconer, J., *Income and Employment, Forestry and Food Security*. Rome: Food and Agriculture Organization of the United Nations, 1987.

Baker, M. B., Jr., DeBano, L. F., Ffolliott, P. F. "Changing Values of Riparian Ecosystems." In *History of Watershed Research in the Central Highlands*, M. B. Baker, Jr., compiler. Fort Collins, CO: Rocky Mountain Research Station, USDA Forest Service, General Technical Report RMRS-GTR-29, 1999, pp. 43-47.

Baker, M. B., Jr., DeBano, L. F., Ffolliott, P. F., Gottfried, G. J. "Riparian-Watershed Linkages in the Southwest." In *Rangeland Management and Water Resources: Proceedings of the American Water Resources Association Specialty Conference*, D. E. Potts, ed. Herndon, VA: American Water Resources Association, TPS-98-1, 1998, pp. 347-355.

Beschta, R. L. "Watershed Management in the Pacific Northwest: The Historical Legacy." In *Land Stewardship in the 21st Century: The Contributions of Watershed Management,* P. F. Ffolliott, M. B. Baker, Jr., C. B. Edminster, M. C. Dillon, and K. L. Mora, tech. coords. Fort Collins, CO: Rocky Mountain Research Station, USDA Forest Service, Proceedings RMRS-P-13, 2000, pp. 109-116.

Bortero, L. S. "Incentives for Community Involvement in Upland Conservation. In *Strategies, Approaches and Systems in Integrated Watershed Management*. Rome: Food and Agriculture Organization of the United Nations, FAO Conservation Guide 14, 1986, pp. 164-172.

Briones, N. D., *Socio-Economics of Watershed Management: The Case of the Lower Agno River Watershed, Luzon, Philippines*. Honolulu, HA: University of Hawaii, PhD Dissertation, 1985.

Bromley, D. W., "Institutions and Institutional Economics in Environmental Policy." In *Environmental Policy Analysis in Developing Countries: Proceedings of a Workshop*. Rosslyn VA: Environmental and Natural Resources Policy and Training Project, U.S. Agency for International Development, 1992, pp. 1-20.

Bromley, D. W., *The Handbook of Environmental Economics*. Oxford, UK: Blackwell Handbooks in Economics, 1995.

Brooks, K. N. Coping with hydro-meteorological disasters: The role of watershed management. Journal of Soil and Water Conservation (Taiwan) 1998; 29:219-231.

Brooks, K. N., Eckman, K. "Global Perspective of Watershed Management." In *Land Stewardship in the 21st Century: The Contributions of Watershed Management,* P. F. Ffolliott, M. B. Baker, Jr., C. B. Edminster, M. C. Dillon, and K. L. Mora, tech. coords. Fort Collins, CO: Rocky Mountain Research Station, USDA Forest Service, Proceedings RMRS-P-13, 2000, pp. 11-20.

Brooks K. N., Ffolliott, P. F. "Watersheds as Management Units: An International Perspective." In *Rangelands in a Sustainable Biosphere: Proceedings of the Fifth International Rangeland Conference*, N. E. West, ed. Denver, CO: Society for Range Management, 1995, pp. 167-169.

Brooks, K. N., Ffolliott, P. F., Gregersen, H. M., DeBano, L. F., *Hydrology and the Management of Watersheds*. Ames, IA: Iowa State University Press, 1997.

Brooks, K. N., Ffolliott, P. F., Gregersen, H. M., Easter, K. W., *Policies for Sustainable Development: The Role of Watershed Management*. Washington, DC: U.S. Department of State, EPAT Policy Brief, 6, 1994.

Brooks, K. N., Gregersen, H. M., Ffolliott, P. F., Tejwani, K. G. "Watershed Management: A Key to Sustainability." In *Managing the Worlds's Forests: Looking for Balance Between Conservation and Development,* N. P. Sharma, ed. Dubuque, IA: Kendall/Hunt Publishing Company, 1992, pp. 455-487.

Brooks, K. N., Gregersen, H. M., Lundgren, A. L., Quinn, R. M., *Manual on Watershed Management Project Planning, Monitoring and Evaluation*. College, Laguna, Philippines: ASEAN-US Watershed Project, 1990.

Brown, T. C., Brown, D., Brinkley, D. Laws and programs for controlling nonpoint source pollution in forest areas. Water Resources Bulletin 1993; 22:1-13.

Cernea, M. M., *Land Tenure Systems and Social Implications of Forestry Development Programs*. Washington, DC: The World Bank, World Bank Staff Working Papers 452, 1981.

Chaney, E., Elmore, W., Platts, W. S., *Livestock Grazing on Western Riparian Areas*. Washington, DC: U.S. Environmental Protection Agency, 1990.

Cheng, J. D., Hsu, H. K., Ho, W. J., Chen, T. C. "Watershed Management for Disaster Mitigation and Sustainable Development in Taiwan." In *Land Stewardship in the 21st Century: The Contributions of Watershed Management,* P. F. Ffolliott, M. B. Baker, Jr., C. B. Edminster, M. C. Dillon, and K. L. Mora, tech. coords. Fort Collins, CO: Rocky Mountain Research Station, USDA Forest Service, Proceedings RMRS-P-13, 2000, pp. 138-148.

Clarke, J. N., McCool, D. C., *Staking Out the Terrain: Power and Performance Among Natural Resource Agencies*. Albany, NY: State University of New York, 1996.

Clary, W., Schmidt, L., DeBano, L. "The Watershed-Riparian Connection: A Recent Concern?" In *Land Stewardship in the 21st Century: The Contributions of Watershed Management,* P. F. Ffolliott, M. B. Baker, Jr., C. B.

Edminster, M. C. Dillon, and K. L. Mora, tech. coords. Fort Collins, CO: Rocky Mountain Research Station, USDA Forest Service, Proceedings RMRS-P-13, 2000, pp. 221-226.

Clary, W. P., Shaw, N. L., Dudley, J. G., Saab, V. A., Kinney, J. W., *Response of a Depleted Sagebrush Steppe Riparian System to Grazing Control and Woody Planting*. Ogden, UT: Intermountain Research Station, USDA Forest Service, Research Paper INT-RP-492, 1996.

Cortner, H. J., Moote, M. A., *The Politics of Ecosystem Management*. Covelo, CA: Island Press, 1999.

Cortner, H. J., Moote, M. A. "Ensuring the Common for the Goose: Implementing Effective Watershed Policies." In *Land Stewardship in the 21st Century: The Contributions of Watershed Management*, P. F. Ffolliott, M. B. Baker, Jr., C. B. Edminster, M. C. Dillon, and K. L. Mora, tech. coords. Fort Collins, CO: Rocky Mountain Research Station, USDA Forest Service, Proceedings RMRS-P-13, 2000, pp. 247-256.

DeBano, L. F., Schmidt, L. J., *Improving Southwestern Riparian Areas Through Watershed Management*. Fort Collins, CO: Rocky Mountain Forest and Range Experiment Station, USDA Forest Service, General Technical Report RM-182, 1989.

DeBano, L. F., Schmidt, L. J. Potential for enhancing riparian habitats in the southwestern United States with watershed practices. Forest Ecology and Management 1990; 33/34:385-403.

DeBano, L. F., Ffolliott, P. F., Brooks, K. N. "Flow of Water and Sediments Through Southwestern Riparian Systems." In *Desired Future Conditions for Southwestern Riparian Ecosystems: Bringing Interests and Concerns Together*, D. W. Shaw and D. M. Finch, tech. coords. Fort Collins CO: Rocky Mountain Forest and Range Experiment Station, USDA Forest Service, General Technical Report RM-GTR-272, 1996, pp. 128-134.

de Steiguer, J. E. "Sustaining Flows of Crucial Watershed Resources." In *Land Stewardship in the 21st Century: The Contributions of Watershed Management*, P. F. Ffolliott, M. B. Baker, Jr., C. B. Edminster, M. C. Dillon, and K. L. Mora, tech. coords. Fort Collins, CO: Rocky Mountain Research Station, USDA Forest Service, Proceedings RMRS-P-13, 2000, pp. 215-220.

Eckman, K., Gregersen, H. M., Lundgren, A. L. "Watershed Management and Sustainable Development: Lessons Learned and Future Directions." In *Land Stewardship in the 21st Century: The Contributions of Watershed Management*, P. F. Ffolliott, M. B. Baker, Jr., C. B. Edminster, M. C. Dillon, and K. L. Mora, tech. coords. Fort Collins, CO: Rocky Mountain Research Station, USDA Forest Service, Proceedings RMRS-P-13, 2000, pp. 37-43.

Ewing, S. Landcare and community-led watershed management in Victoria, Australia. Journal of the American Water Resources Association 1999; 35:663-673.

Federal Interagency Stream Restoration Working Group, *Stream Corridor Restoration: Principles, Processes, and Practices*. Washington, DC: Government Printing Office, 1998.

Ffolliott, P. F., Brooks, K. N. "Multiple Use: Achieving Diversified and Increased Income within a Watershed Management Framework." In *Strategies, Approaches, and Systems in Integrated Watershed Management*. Rome: Food and Agriculture Organization of the United Nations, FAO Conservation Guide 14, 1986, pp. 114-123.

Ffolliott, P. F., Brooks, K. N. Knowledge of physical and biological relationships for economic evaluations: A key to sustainable watershed management. Proceedings of the Watershed Conservation and Sustainable Management Seminar; 1998 April 22-23; Taichung, Taiwan; Taichung, Taiwan: Soil and Water Conservation Department, National Chung-Hsing University, 1998, pp. 1-13.

Ffolliott, P. F., Moshe, I., Sammis, T. W. "A Planning Process." In *Arid Lands Management: Toward Ecologicall Sustainability*, T. W. Hoekstra and M. Shachak, eds. Urbana, Il: University of Illinois Press, 1999, pp. 171-178.

Fields, B. C., *Environmental Economics: An Introduction*. New York: McGraw-Hill, Inc., 1994.

Frasier, G. W., Hart, R. H., Schuman, G. E. "Impacts of Grazing Intensity on Infiltration/Runoff Characteristics of a Shortgrass Prairie. In *Rangelands in a Sustainable Biosphere: Proceedings of the Fifth International Rangeland Conference*, N. E. West, ed. Denver, CO: Society for Range Management, 1995, pp. 159-160.

Garrett, H. E., Rietveld, W. J., Fisher, R. F., eds., *North American Agroforestry: An Integrated Science and Practices*. Madison, WI: American Society of Agronomy, Inc., 2000.

Gelt, J. "Watershed Management: A Concept Evolving to Meet New Needs." In *Land Stewardship in the 21st Century: The Contributions of Watershed Management*, P. F. Ffolliott, M. B. Baker, Jr., C. B. Edminster, M. C. Dillon, and K. L. Mora, tech. coords. Fort Collins, CO: Rocky Mountain Research Station, USDA Forest Service, Proceedings RMRS-P-13, 2000, pp. 65-73.

Gordon, A. M., Newman, S. M., eds., *Temperate Agroforestry Systems*. London, UK: CAB International, 1997.

Gregersen, H. M. People, trees, and rural development: The role of social forestry. Journal of Forestry 1988; 86(10):22-30.

Gregersen, H. M., Contreras, A., *Economic Assessment of Forestry Impacts*. Rome, Italy: Food and Agriculture Organization of the United Nations, FAO Forestry Paper 106, 1992.

Gregersen, H. M., McGaughey, S. E., *Improving Policies and Financing Mechanisms for Forestry Development.* Washington, DC: Economic and Social Department, Inter-American Development Bank, 1985.

Gregersen, H. M., Draper, S., Elz, D., eds., *People and Trees: The Role of Social Forestry in Sustainable Development.* Washington, DC: Economic Development Institute of The World Bank, 1989.

Gregersen, H. M., Brooks, K. N., Dixon, J. A., Hamilton, L. S., *Guidelines for Economic Appraisal of Watershed Management Projects.* Rome: Food and Agriculture Organization of the United Nations, FAO Conservation Guide 16, 1987.

Hamilton, L. S. "The Watershed as a Management Unit and a Conceptual Framework for Integrated Watershed Management." In *Integrated Watershed Management Research for Developing Countries*, K. W. Easter, M. M. Hufschmidt, eds. Honolulu, HA: Environment and Policy Institute, East-West Center, 1985, pp. 24-25.

Heede, B. H., *Stream Dynamics: An Overview for Land Managers.* Fort Collins, CO: Rocky Mountain Forest and Range Experiment Station, USDA Forest Service, General Technical Report RM-72, 1980.

Heede, B. H., Harvey, M. D., Laird, J. R. Sediment delivery linkages in a chaparral watershed following a wildfire. Environmental Management 1988; 12:349-358.

Ilhardt, B. L., Verry,, E. S., Palik, B. J. "Defining Riparian Areas." In *Riparian Management in Forests of the Continental Eastern United States*, E. S. Verry, J. W. Hornbeck, and C. A. Dolloff, eds. Boca Raton, FL: Lewis Publishers, 2000, pp. 23-42.

International Wildlife. Most states ignore leading causes of water pollution, NFW finds. International Wildlife (July/August) 2000. p. 6.

Ingles, A. W., Musch, A., Quist-Hoffmann, H., *The Participatory Process for Supporting Collaborative Management of Natural Resources: An Overview.* Rome: Food and Agriculture Organization of the United Nations, 1999.

Krueper, D. J. "Effects of Livestock Management on Southwestern Riparian Ecosystems." In *Desired Future Conditions for Southwestern Riparian Ecosystems: Bringing Interests and Concerns Together*, D. W. Shaw and D. M. Finch, D. M., tech. coords. Fort Collins, CO: Rocky Mountain Forest and Range Experiment Station, USDA Forest Service, General Technical Report RM-GTR-272, 1996, pp. 281-301.

Krutilla, J. A., Bowes, M. D., *Multiple Use Management: The Economics of Public Forest Lands.* Washington, DC: Resources for the Future, 1989.

Lant, C. L. Introduction - Human dimensions of watershed management. Journal of the American Water Resources Association 1999; 35:483-486.

LaFayette, R. A., Pruitt, J. R., Zeedyk, W. D., *Riparian Area Enhancement Through Road Management.* Albuquerque, NM: Southwestern Region, USDA Forest Service, 1992.

Loomis, J. B., *Integrated Public Lands Management.* New York: Columbia University Press, 1993.

Long, J. W. "Restoration of Gooseberry Creek." In *Land Stewardship in the 21st Century: The Contributions of Watershed Management,* P. F. Ffolliott, M. B. Baker, Jr., C. B. Edminster, M. C. Dillon, and K. L. Mora, tech. coords. Fort Collins, CO: Rocky Mountain Research Station, USDA Forest Service, Proceedings RMRS-P-13, 2000, pp. 227-233.

Lopes, V. L., Ffolliott, P. F., Fogel, M. M. "Integrated Watershed Management for Sustainable Use of Natural Resources: A Framework for Consideration." In *First International Seminar of Watershed Management: Proceedings*, J. Castillo, M. Tiscareno, and I. Sanchez Cohen, eds. Hermosillo, Sonora, Mexico; Tucson, AZ: Universidad de Sonora-University of Arizona, 1993, pp. 105-114.

Lynch, J. A., Corbett, E. S., Mussallem, K. Best Management Practices for controlling nonpoint-source pollution on forested watersheds. Journal of Soil and Water Conservation 1985; 40:164-167.

MacDicken, K. G., Vergara, N. T., *Agroforestry: Classification and Management.* New York: John Wiley & Sons, Inc., 1990.

Medina, A. L. "Native Aquatic Plants and Ecological Condition of Southwestern Wetlands and Riparian Areas." In *Desired Future Conditions for Southwestern Riparian Ecosystems: Bringing Interests and Concerns Together*, D. W. Shaw and D. M. Finch, tech. coords. Fort Collins, CO: Rocky Mountain Forest and Range Experiment Station, USDA Forest Service, General Technical Report RM-GTR-272, 1996, pp. 329-335.

Minckley, W. L., Rinne, J. M. Large woody debris in hot-desert streams: An historical review. Desert Plants 1985; 7:142-153.

Moore, E., James, E., Kinsinger, F., Pitney, K., Sainsbury, J., *Livestock Grazing Management and Water Quality Protection.* Washington, DC: U.S. Environmental Protection Agency, EPA 910/9-79-67. 1979.

Morgan, J. R. "Watersheds as Functional Regions: A Case Study of the Hawaiian Ahupuaa." In *Integrated Watershed Management Research for Developing Countries*, K. W. Easter, M. M. Hufschmidt, eds. Honolulu, HA: Environment and Policy Institute, East-West Center, 1985, pp. 25-26.

Morgan, J. R. "Watersheds in Hawaii: An Historical Example of Integrated Management." In *Watershed Resources Management: An integrated Framework with Studies from Asia and the Pacific*, K. W. Easter, Dixon, J. A., Hufschmidt, eds. Boulder, CO: Westview Press, 1986, pp. 133-144.

Nair, P. K. R., *An Introduction to Agroforestry*. Dordrecht, The Netherlands: Kluwer Academic Press, 1993.

National Research Council, *New Strategies for America's Watersheds*. Washington, DC: National Research Council, National Academy Press, 1999.

Neary, D. G. "Changing Perceptions of Watershed Management from a Retrospective Viewpoint." In *Land Stewardship in the 21st Century: The Contributions of Watershed Management,* P. F. Ffolliott, M. B. Baker, Jr., C. B. Edminster, M. C. Dillon, and K. L. Mora, tech. coords. Fort Collins, CO: Rocky Mountain Research Station, USDA Forest Service, Proceedings RMRS-P-13, 2000, pp. 167-176 .

Porto, M., Porto, R. L. L., Azevedo, L. G. T. A participatory approach to watershed management: The Brazilian system. Journal of the American Water Resources Association 1999; 35:675-683.

Quinn, R. M., Brooks, K. N., Ffolliott, P. F., Gregersen, H. M., Lundgren, A. L., *Reducing Resource Degradation: Designing Policy for Effective Watershed Management*. Washington, DC: U.S. Department of State, EPAT Working Paper 22, 1995.

Raintree, J. B., ed., *Land, Trees and Tenure: Proceedings of an Integrated Workshop on Tenure Issues in Agroforestry*. Madison, WI: Land Tenure Center, University of Wisconsin, 1987.

Reid, L. M., *Research and Cumulative Watershed Effects*. Berkeley, CA: Pacific Southwest Forest and Range Experiment Station, USDA Forest Service, General Technical Report PSW-GTR-141, 1993.

Schlutz, R. C., Colletti, J. P., Isenhart, T. M., Marquez, C. O., Simpkins, W. W., Ball, C. J. "Riparian Forest Buffer Practices." In *North American Agroforestry: An Integrated Science and Practices*, H. E. Garrett, Rietveld, W. J., Fisher, R. F., eds., Madison, WI: American Society of Agronomy, Inc., 2000, pp. 189-281.

Sidle, R. C. "Watershed Challenges for the 21st Century: A Global Perspective for Mountainous Terrain." In *Land Stewardship in the 21st Century: The Contributions of Watershed Management,* P. F. Ffolliott, M. B. Baker, Jr., C. B. Edminster, M. C. Dillon, and K. L. Mora, tech. coords. Fort Collins, CO: Rocky Mountain Research Station, USDA Forest Service, Proceedings RMRS-P-13, 2000, pp. 45-56.

Somerville, M., DeSimone, D. "Securing Clean Water: A Secret to Success." In *Land Stewardship in the 21st Century: The Contributions of Watershed Management,* P. F. Ffolliott, M. B. Baker, Jr., C. B. Edminster, M. C. Dillon, and K. L. Mora, tech. coords. Fort Collins, CO: Rocky Mountain Research Station, USDA Forest Service, Proceedings RMRS-P-13, 2000, pp. 213-214.

Thorud, D. B., Brown, G. W., Boyle, B. J., Ryan, C. M. "Watershed Management in the United States in the 21st Century." In *Land Stewardship in the 21st Century: The Contributions of Watershed Management,* P. F. Ffolliott, M. B. Baker, Jr., C. B. Edminster, M. C. Dillon, and K. L. Mora, tech. coords. Fort Collins, CO: Rocky Mountain Research Station, USDA Forest Service, Proceedings RMRS-P-13, 2000, pp. 57-64.

Tietenberg, T., *Environmental and Natural Resources Economics*. Glenview, IL: Scott Foresman and Co., 1988.

Tillman, G., *Environmentally Sound Small-Scale Water Projects: Guidelines for Planning*. New York: Codel Publication, 1981.

Towns, E. S. "Resource Integration and Shared Outcomes at the Watershed Scale." In *Land Stewardship in the 21st Century: The Contributions of Watershed Management,* P. F. Ffolliott, M. B. Baker, Jr., C. B. Edminster, M. C. Dillon, and K. L. Mora, tech. coords. Fort Collins, CO: Rocky Mountain Research Station, USDA Forest Service, Proceedings RMRS-P-13, 2000, pp. 74-80.

Tracy, J. C., Bernknopf, R., Forney, W., Hill, K. "A prototype for understanding the effects of TMDL standards: Values to sediment loads in the Lake Tahoe Basin." In *Watershed 2000: Science and Engineering Technology for the New Millennium*, M. Flug and D. Frevert, eds. Reston, VA: American Society of Civil Engineers, 2000. [CD-ROM] Windows.

U.S. Environmental Protection Agency, *Water Quality Strategy Paper*. Washington, DC: U.S. Environmental Protection Agency, 1974.

U.S. Environmental Protection Agency, *Clean Water Action Plan: Restoring and Protecting America's Water*. Washington, DC: U.S. Environmental Protection Agency, Internet WWW page http:/www.cleanwater.gov/, January 23, 1999.

Verry, E. S., Hornbeck, J. W., Dolloff, A. C., eds., *Riparian Management in Forests of the Continental Eastern United States*. Boca Raton, FL: Lewis Publishers, 2000a.

Verry, E. S., Hornbeck, J. W., Todd, A. H. "Watershed Management in the Lake States and Northeastern United States." In *Land Stewardship in the 21st Century: The Contributions of Watershed Management,* P. F. Ffolliott, M. B. Baker, Jr., C. B. Edminster, M. C. Dillon, and K. L. Mora, tech. coords. Fort Collins, CO: Rocky Mountain Research Station, USDA Forest Service, Proceedings RMRS-P-13, 2000b, pp. 81-92.

Whiting, P. J., Stamm, J. F., Moog, D. R., Orndorff, R. L. Sediment-transport flows in headwater streams. Bulletin of the Geological Society of America 1999; 111:450-466.

World Resources Institute, *World Resources: A Guide to the Global Environment*. New York: Oxford University Press, 1996.

6

FUTURE PROTOCOLS

Society will likely implement protocols when adopting a watershed management approach to future land stewardship. Three of the protocols, identified by the participants at the conference on "Land Stewardship in the 21st Century: The Contributions of Watershed Management" as high priority (Ffolliott et al., 2000a), are considered in this chapter. They are:

- Anticipating future watershed conditions and maintaining landscape integrity;
- Responding to increased demands for land, water, and other natural resources; and
- Implementing appropriate land-use policies to attain these protocols and meet other demands for natural and economic resources.

Water unifies watershed management protocols. Implementation of effective land-use policies that incorporate ecological understanding and promote democratic ideals is also necessary. Guidelines to achieve this are the immediate integration of the political process, building bridges to citizens, examining laws, rights, and responsibilities, strengthening administrative capacity, and looking beyond the watershed to a broader scale. Before examining these future protocols, we reinforce the rationale for a watershed management approach to land stewardship.

6.1. RATIONALE FOR A WATERSHED MANAGEMENT APPROACH TO LAND STEWARDSHIP

Natural ecosystems are sometimes managed to improve the welfare of people using the practices, projects, and programs of a watershed management approach to land stewardship. Watershed management activities should maximize the sustainable use of natural resources and the equitable distribution of benefits and costs, while minimizing social disruption and adverse environmental impacts (Gregersen et al., 1987; Brooks et al., 1990, 1992, 1994, 1997; Easter et al., 1995; de Steiguer, 2000). A watershed highlights the physical aspects of a landscape that, if unrecognized, can result in the loss of natural resources, degradation of ecological functioning and environmental conditions, and problems implementing watershed management interventions. The watershed as a planning and management unit also contains the relationships and issues that must be considered in future land stewardship.

The rationale for a watershed management approach to land stewardship, whether the focus is on water resources, forestry activities, livestock production, agriculture, or combinations of these land uses, has been summarized by Easter and Hufschmidt (1985), Brooks et al., (1990; 1992), Quinn et al. (1995), and others as follows:

- A watershed or river basin is a functional area of a landscape that includes the key interrelationships and interdependencies of concern for land, water, and other natural resource management.
- A watershed management approach is a logical basis for evaluating the biophysical relationship between upstream and downstream activities. This approach is also holistic, enabling planners, managers, and decision-makers to identify and evaluate all of the relevant facets of effective land stewardship including on- and off-site changes and impacts. The approach accounts for the whole complex of biophysical, economic, social, and institutional factors that affect development of sustainable management programs.
- Using a watershed as a management unit is economically logical since it internalizes many off-site effects involved with land stewardship.
- A watershed management approach allows for assessment of environmental impacts including the effects of land-use activities on large and small upstream and downstream ecosystems. Consequently, the effects of upland disturbances, which often result in a chain of downstream consequences, can be readily examined and evaluated within a watershed context.
- Using the watershed as a management unit allows for the consideration and evaluation of human interactions with the environment.
- A watershed management approach to land stewardship can be effectively and efficiently integrated into other natural resource conservation and development programs—for example, soil and water conservation, forestry, farming systems, and rural and community development.

A watershed management approach considers natural ecosystems and social systems. In many respects, a watershed management approach contrasts with earlier approaches to land stewardship. Previously, land stewardship had two perspectives—either focusing on natural ecosystems and ignoring the social systems or focusing on social systems and considering the natural ecosystem a constraint (Easter and Hufschmidt, 1985; Gregersen et al., 1987; Quinn et al., 1995; Brooks et al., 1997). A question often asked from the first perspective is, "How can people be removed to preserve the integrity and functioning of a natural ecosystem?" From the second perspective, "What must be done to the natural ecosystem to extract more services and income for the social system?" is a typical inquiry. A watershed management approach to land stewardship effectively answers both questions.

6.2. PROTOCOLS

Embedded in the three protocols—anticipating future landscape conditions, responding to increased demands for water, and implementing effective policies and policy instruments—which are considered in this chapter, are other protocols that relate to watershed management contributions to land stewardship that were discussed in Chapter 5. Further, we suggest that the three protocols encompass many other contributions of a watershed management approach that were not explicitly considered.

6.2.1. Anticipating Future Landscape Conditions

Anticipating landscape conditions in the 21st century is a difficult task. Different people have different perceptions of what future landscapes should look like (Box 6.1). However, nearly everyone agrees that however the landscapes appear, they should be managed to recognize and sustain the landscape's inherent ecological properties, the composition and functioning of the ecosystems present, and society's perceptions of how the landscapes should be used. This view is the ecosystem approach to landscape management (Salwasser et al., 1993; Jones et al., 1995; Brown and Marshall, 1996; Webster and Chappelle, 1997). We believe that a watershed management approach to future land stewardship is consistent with an ecosystem approach to landscape management.

6.2.1.1. Relationship of Multiple-use to Ecosystem Management

The multiple-use philosophy of natural resource management served society well into the 1980s. Lands were characterized largely on their capacities to provide commodities and amenities. Research was discovering the factors that limited the realization of these capacities, while management was reducing or removing the limitations (Kessler et al., 1992; Ffolliott et al., 1995). Answers to questions about natural resources required identification of optimal yields among the desired but often competing resources. However, multiple-use management is not necessarily the best approach to current land management.

People began to ask questions in the 1990s about how to balance a wide range of potential natural resource uses and values on a landscape. In responding to these questions, the National Research Council (1990) of the United States, proposed an alternative paradigm—the ecosystem approach to landscape management. The new model, broadened the earlier multiple-use philosophy to one of holistically conceived ecosystem management (Kessler et al., 1992; Swank and VanLear, 1992). This approach is applicable to natural resource management on all landscapes, and it requires natural resource managers to view landscapes as a comprehensive context of living systems, which includes soils, plants, animals, minerals, climate, topography, and people. The ecosystem approach to natural resource management is important beyond the traditional commodity and amenity uses of the landscape, which

Box 6.1 Anticipating Future Landscape Conditions: A Case Study

A group of ranchers in the Malpai Borderland Region of the Southwestern United States work with governmental agencies, universities, and environmental groups to reduce the threat of regional property and ecosystem fragmentation (Gottfried et al., 1999, 2000). The region contains a variety of ecosystems extending from low-elevation desert shrub and tobosa grasslands to high-elevation ponderosa pine and Douglas-fir forests. The mountains and valleys are home to diverse populations of plants and wildlife including some species rarely found in the United States. Landownership is divided between private individuals and federal and state agencies. The Malpai Borderlands Region is home to a viable ranching community (McDonald, 1995, 2000). Property and habitat fragmentation, which is obvious in many adjacent valleys, has not reached the area, largely due to the efforts of the Malpai Borderlands ranchers, and their collaborators, who are working diligently to ensure that it never does.

optimized the production or use of one or a few natural resources and often compromised the balances, values, and functional properties of the whole ecosystem. Additionally, production or use of individual natural resources, as previously practiced, led to situations of non-sustainability.

The ecosystem approach relates the ecological and social conditions of the area to the land, levels of land uses, and flows of crucial natural resources that are compatible with these conditions (Kessler et al., 1992; Irland, 1994). The relations reflect the goals and objectives satisfied through area-oriented multiple-use management in which information obtained from resource-oriented multiple-use management is applied to the land area's capability and suitability to produce the resource (Ridd, 1965; Brooks et al., 1997). Area-oriented multiple-use management uses the information needed to describe the potential for natural resource development obtained from resource-oriented multiple-use management[36], and relates it to local, regional, and national supplies of and demands for natural resources.

6.2.1.2. Watershed Basis to Landscape Management

Ecosystem management involves concepts, principles, and applications that evolve and adapt with science, economics, and demographics (Staebler, 1994). Some views of ecosystem management are discipline oriented, others consider political boundaries, and still others contemplate land ownership. Whatever the perception, the ecosystem approach to achieving future landscape conditions can be effectively and efficiently applied on a watershed basis (Figure 6.1). Therefore, a watershed management approach to achieve specified landscape conditions is a logical way to manage landscapes to meet future conditions wanted by society. That is, the essentials of ecosystem management (Brooks and Grant, 1992; Swank and VanLear, 1992; Miadenoff and Pastor, 1993; Irland, 1994) and a watershed approach to achieving a desired landscape condition are similar in concept and application. These similarities include:

- The essence of ecosystem management and a watershed management approach to land stewardship is their objectives and wider spatial and temporal scales—not necessarily a set of particular management practices implemented to achieve those objectives. Protection and enhancement of ecosystem integrity are essential to both approaches.
- Both ecosystem management and a watershed management approach are intensive forms of managing land, water, and other natural resources to maintain environmental quality and standards. Their implementation requires concentrated planning and coordination, spatially detailed data sets, and sophisticated land-use management prescriptions. The activities are expected to produce a productive combination of resource benefits over time.
- Flows of clean water and other natural and economic resources are sustained within the capacities of the ecological productivity limits of the managed landscapes.
- Landscape traits, including the protection of waterways, identification and protection of critical habitat components, avoidance of ecosystem fragmentation, and connectivity, are emphasized in ecosystem management and a watershed management approach to land stewardship.

[36] Management that focuses on the production capabilities of natural resources by considering the land's capabilities and suitability.

- Cumulative effects of ecosystem management and a watershed management approach are a basis for monitoring and evaluating their effectiveness in meeting the specified landscape conditions.
- Many think that ecosystem management and a watershed management approach to land stewardship ensures against the ecological, economic, and social effects of future global climate change and other long-term changes in ecosystem functioning.

While anticipating what future landscape conditions society might want is difficult, a level of flexibility is achieved through a watershed management approach to land stewardship. However, caution is warranted. Recognizing that some management practices implemented to attain future landscape conditions could require sweeping modifications of vegetation and other ecosystem properties is important. Such management practices could jeopardize other resource values in the landscape and be largely irrevocable, at least in the short-run, because they are easily implemented but not be easily undone. Careful participatory planning efforts are necessary to ensure that this situation does not occur.

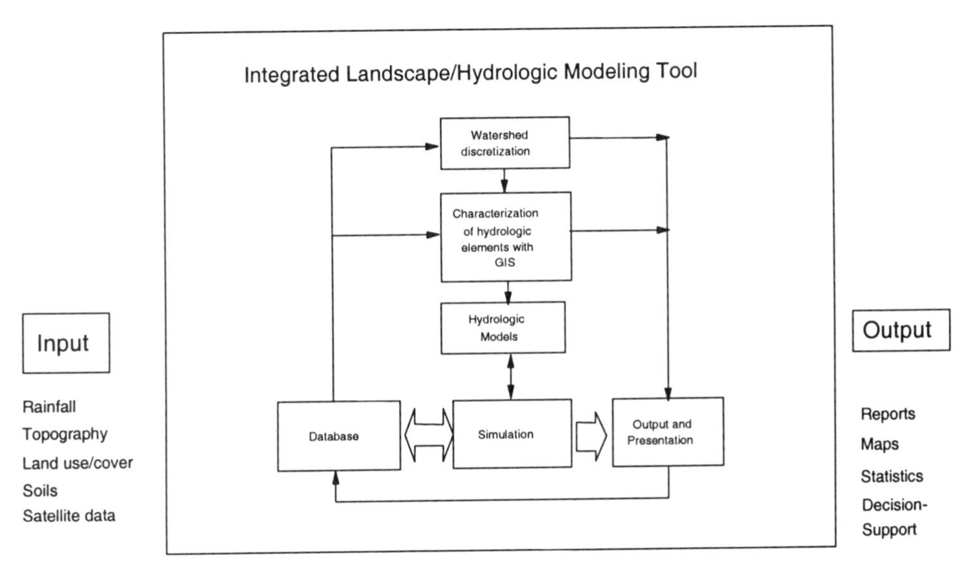

Figure 6.1. A landscape-watershed modeling framework encompassing a database, simulation capabilities, and a user interface (from Hernandez and Miller, 2000).

6.2.2. Responding to Increased Water Demands

Other chapters have established that there are rapidly growing demands for land, water, and other natural resources throughout the world, and that the sustainability of resource flows is expected to remain important to many countries. While this book has dealt broadly with watershed management practices, projects, and programs and the multiple outputs obtained from these efforts, we focus below on watershed management in relation to increasing water demands. We take this perspective (Gregersen et al., 2000) because:

- We believe that water will be the key watershed management issue in the 21st century;
- Water is the single unifying theme that provides cohesion to the varying elements of integrated watershed management; and
- Water is the best example of why addressing only the supply side of natural resource management is insufficient when responding to the increasing demands for natural resources. Planners, managers, and decision-makers must also consider the demand side.

6.2.2.1. Issues

While the increasing cost of water will likely reduce its level of use, the increasing scarcities of clean water will remain (Gregersen et al., 2000). Many issues that will probably demand attention were previously detailed in this book. However, broadly speaking, the following trends are evident regarding the future needs of water and watershed management in relation to meeting increasing demands for high-quality water.

- Accumulating evidence shows that watershed management researchers have been successful in developing the technical means to secure the most and best water available at a given time and place. While continuing technical research on water resources will be necessary, the focus in the 21st century must be on management and institutional protocols necessary to effectively use present technologies (Lant, 1999; Gregersen et al., 2000).
- There is increasing evidence that people and institutions are responding to floods, drought, and other crises by reducing water consumption and increasing investments in the infrastructures needed to sustain high-quality water supplies. However, most of the emphasis has been on crisis management rather than on developing the prevention strategies necessary to avoid unsustainable use and development of watershed resources, and the future crisis that their unavailability will cause (Whilhite, 1997; Gregersen et al., 2000).
- We are moving toward decentralization of responsibilities for the environment and participatory management of watersheds and water resources. Locally-led watershed management initiatives in the United States (Lant, 1999) are increasing in number, and participatory management of natural resources is growing worldwide (see Chapter 3).
- Increasing importance in the globalization of environmental issues and responses is occurring. Recently, there has been a growing international trade of commodities produced from natural resources and a proliferation of environmental and natural resource international conventions and programs—for example, agreements on climate change, biodiversity, desertification, and fisheries.

Society must develop effective combinations of local, national, and global institutional arrangements in response to the mounting scarcities of water and other natural resources (Brooks et al., 1994; Quinn et al., 1995; Gregersen et al., 2000). These responses should:

- Take advantage of the technical knowledge about how to manage water and other natural resources;

- Use information from and the understanding of the local users of watershed resources regarding the relevant issues and options associated with a watershed management approach to land stewardship; and
- Help resolve potential upstream and downstream user conflicts at different locations on a watershed or within a large river basin.

6.2.2.2. Recent Responses to Water Issues

Participants at the International Conference on Water and the Environment, held in Dublin, Ireland, on January 26 through 31, 1992, adopted the *Dublin Conference and the Conference Report* to help guide development of effective water policies worldwide (The World Bank, 1993; Hydrology and Water Resources Programme of the World Meteorological Organization, 1999; Gregersen et al., 2000). The water-policy agenda contained in this report is consistent with a watershed management approach to land stewardship. The 4 principles points presented in the report are that:

- Water is a finite and vulnerable resource that is essential to sustaining life, economic development, and the environment;
- Water development and management should be based on a participatory approach involving users, planners, and policymakers at all levels;
- Women should have a central role in providing, managing, and safeguarding water; and
- Water has an economic value and should be recognized as an economic goal.

The 10 policy agendas adopted by the Dublin Conference participants are:

- Alleviate poverty and disease through water resource development and management to accelerate provisions for food, water, and sanitation to the quarter of the world's population that lacks these basic needs. Participants felt that by the end of the 1990s, food scarcity was not a production problem, but instead was due to a lack of income and effective distribution.
- Protect against floods, droughts, and other natural disasters by increasing investments in basic hydrologic data collections to predict their recurrence intervals better and upgrade disaster preparedness.
- Contribute to water conservation and reuse through improved agricultural irrigation practices and recycling. Holistic management of existing water resources could reduce costly losses.
- Provide for sustainable urban development, which the curtailment of affordable water currently threatens. The generation of future reliable water supplies must be based on the appropriate water charge and discharge controls.
- Contribute to agricultural production and rural water supplies by applying water-saving technologies and management methods that help communities introduce institutions and incentives for the rural population to adopt new water-conservation approaches. In many countries, obtaining sufficient amounts of food remains a high priority and, therefore, agriculture must meet the food requirements for increasing populations, while conserving water for other uses.

- Protect aquatic ecosystems from the disruption of water flows and from pollution discharge that requires expensive water treatments, destroys aquatic flora and fauna, and limits recreational opportunities. Integrated management of watersheds and river basins—a watershed management approach—safeguards aquatic ecosystems and produces benefits that are available to society on a sustainable basis.
- Resolve water conflicts by reconciling the interests of all involved stakeholders, monitoring water quantity and quality, developing agreed upon action plans, exchanging pertinent information, and enforcing agreements. Prepare and implement integrated river-basin-management plans that all affected governments and stakeholders can endorse.
- Invest in institutions responsible for and people assigned to implement programs for water and sustainable development through capital investments and to build the capacity of the institutions and the people to plan and implement these programs.
- Enhance the knowledge base necessary for effective water management through more accurate measurements of water cycle components and through other environmental characteristics that effect water delivery. Users can understand the data and their application through interdisciplinary research, monitoring, and analytical techniques.
- Improve personnel, institutional, and legal arrangements by identifying training necessary to plan and implement effective water-resources-management programs and by providing the required training and working conditions to retain the trained personnel. This policy-agenda item requires authorized institutional and legal arrangements, including those for effective water-demand management.

The Dublin Conference formed the basis for the Global Partnership on Water developed to accommodate the recommendations of participants at the United Nations Conference on the Environment and Development (UNCED), held in Rio de Janeiro in June 1992. Participants recommended more effective policies for water and the sustainable development of natural resources (Gregersen et al., 2000). The objectives of this partnership are to:

- Support integrated water-resource-management programs between governments and existing networks by forging new collaborative arrangements;
- Encourage government, aid, and donor agencies, and other stakeholders to adopt consistent and mutually complementary policies and programs;
- Build mechanisms for sharing relevant information and experiences;
- Develop innovative, effective solutions to problems common to integrated water resource management;
- Suggest practical policies and sound management practices; and
- Help match identified needs to available resources.

6.2.2.3. Future Responses to Water Issues

Implications of the Dublin Conference, UNCED, and similar policy-oriented gatherings are clear. Significant future water increases are needed to meet anticipated demands and avoid water shortages in many countries. These efforts will require that reductions in the per capita water consumption (demand side) and development of new and improved water supplies (supply side) be considered (Ffolliott et al., 2000b; Gregerson et al., 2000). Greater use efficiency and reduced per capita water consumption is achieved by improving water-use efficiency. Developing new and improved water supplies can increase the supplies of usable water. Several options exist within each of these actions that planners, managers, and decision-makers can consider when responding to increased water demands (Figure 6.2). These options are a watershed management approach to land stewardship.

Improving management of current water supplies is necessary despite any progress made to reduce water demands or increase water supplies (Brooks et al., 1994; Quinn et al., 1995; Ffolliott et al., 2000b; Gregersen et al., 2000). Effective technology applications must be encouraged, while increased public awareness about the importance of balancing the economic and environmental values of available watershed resources is essential. Implementing effective policies to guide societal responses to increased water demands and other issues related to ensuring sustainable supplies of high-quality water are also necessary (see below).

6.2.3. Implementing Effective Policies and Policy Instruments

In many countries, policies exist to guide the effective implementation of the above responses to the increasing demand for land, water, and other natural resources. A watershed management approach to land stewardship efficiently applies, monitors, and enforces these policies for sustainable development (Brooks et al., 1994, 1997; Quinn et al., 1995; Geregersen et al., 2000). To encourage stakeholders to adopt the policies, one or more of the following instruments is useful:

- Regulatory mechanisms such as regulations, land and water rights, prohibitions, and licensing;
- Fiscal and financial instruments such as prices, taxes, subsidies, and fines; and
- Public investment, such as technical assistance, education, structure installation, infrastructures, and more responsive and effective land stewardship.

Promotion of local commitment and participation, within the existing social, political, and institutional setting, also contributes to successful policy implementation (Gregersen et al., 2000). The setting is unique for each locale, region, and country that relates to the organizations, customs and rights, and laws, regulations, and informal rules that influence the success of a policy or action.

6.2.3.1. Necessary Institutional Arrangements

Effective implementation of a policy can require modification to existing institutional arrangements or developing new policy instruments. Institutional arrangements specify who benefits from the land, water, and other natural resource management. To manage the resource use effectively, well designed, efficiently functioning institutional arrangements should establish land and water rights, regulations, pricing mechanisms, and governmental interventions. Logically, inadequate institutional arrangements impede efficient resource use. Institutional arrangements also build an interface between the government and private sectors in land, water, and natural resource management (Gregersen et al., 1994, 2000; Cortner and Moote, 2000). Watershed and all other natural resource management usually involve a combination of government and private sector activity. Once this association has been decided, the policy instruments that most effectively accomplish the objectives and goals are selected. Some combination of policy actions and instruments is usually more effective than a single action or instrument (Box 6.2).

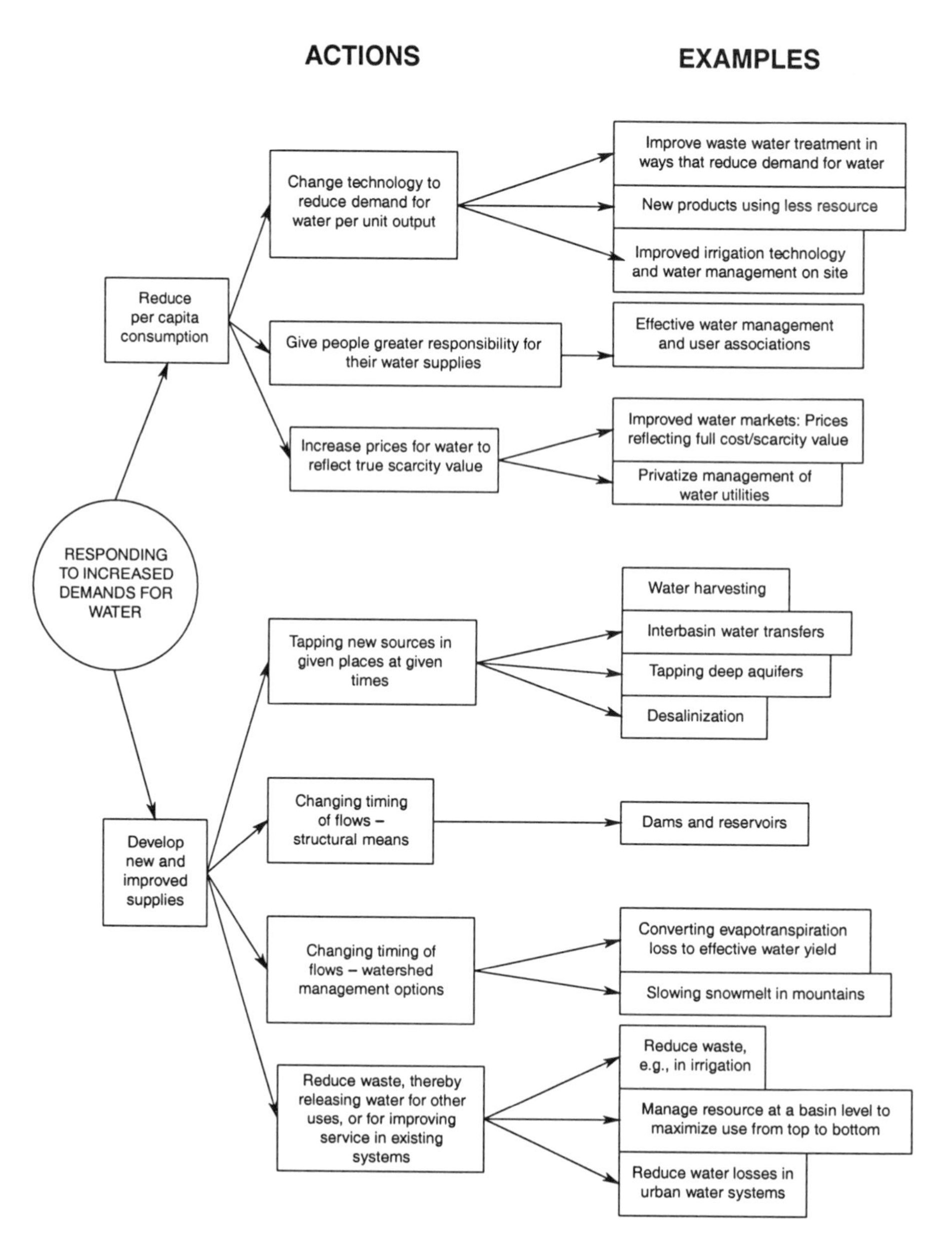

Figure 6.2. Opinions for responding to increased demands for water (adapted from Gregerson et al., 2000).

6.2.3.2. *Increasing Public Awareness*

The public's increasing environmental awareness and concern about the world condition for future generations (Brooks et al., 1994; Quinn et al., 1995; Gregersen et al., 2000) translates into increased political consciousness and calls for action. The growing level of environmental awareness also helps create effective, acceptable distribution of the benefits and costs of watershed management programs, which is the ultimate goal of these programs. A watershed management approach provides a comprehensive and practical biophysical and institutional framework for policy development (see Chapter 5) and implementation and, more generally, for solving complex natural resources problems. However, barriers to the adoption of a watershed management approach to land stewardship must be overcome for this framework to be useful.

Box 6.2. Water Resource Problems and Policy Actions: Some Examples

Policy actions and related policy instruments used to help resolve water problems and to improve water resource management is presented below (Easter et al., 1995). Not all policy actions will be appropriate for each situation and those taken to solve a problem will probably affect others. A combination of policy instruments is usually more effective than a single instrument to establish a policy for a water resource problem.

Policy Instrument	Variable Water Supplies	Water Shortages	Declining Water Quality
1. Regulatory mechanisms and institutions	a. Use floodplain zoning	a. Establish water-user associations	a. Set water-quality standards
	b. Establish priorities for water use in droughts	b. Establish river-basin entities	b. Establish land-use zoning around streams and within watersheds
		c. Impose restrictions on use	
2. Fiscal incentives	a. Use peak-load pricing	a. Use opportunity-cost pricing	a. Make polluter pay for damages
	b. Legalize water markets	b. Legalize water markets	b. Use tradable pollution permits
3. Direct public investments	a. Public groundwater development	a. Transfer water from surplus regions	a. Install waste-treatment plants
	b. Expand reservoir capacity	b. Install water meters	b. Install aerators on polluted rivers
		c. Install water-saving toilets	

6.3. ADOPTION OF A WATERSHED MANAGEMENT APPROACH TO LAND STEWARDSHIP

Some barriers to the adoption of a watershed management approach to future land stewardship exist, while others are thought to exist. Either way, the barriers are overcome through the combined efforts of technically-trained planners and managers, decision-makers, locally-led advocacy groups, and other concerned stakeholders.

6.3.1. Barriers

The basic concepts underlying a watershed management approach to land stewardship also explain why this approach has not previously been widely adopted (Gregersen et al., 1987; Working Group on Watershed Management and Development, 1988; Brooks et al., 1990, 1992; Quinn et al., 1995). Watershed management practices implemented by one political unit, often affect those living outside that political unit. Therefore, little incentive exists for consideration of any watershed management approach that could create a conflict situation.

A question commonly asked by upstream land users demonstrates another barrier to the adoption of watershed management approach, "Why should we accomplish or support watershed management practices that will primarily benefit downstream users?" The response comes as another question, "Why should decision-makers expect upstream land users to accomplish or support watershed management practices that will primarily benefit downstream users, if they are not compensated for the costs of such activities?" The inequity of who pays for and who benefits from a watershed management practice must be resolved to gain the support necessary for adoption of a watershed management rationale to land stewardship. Other often cited barriers to a watershed management approach to land stewardship include:

- A lack of awareness or understanding about the concepts and practices of watershed management by the planners, managers, and decision-makers responsible for land stewardship. Additionally, the public has often limited watershed management practices due to lack of understanding.
- Technically-trained watershed managers have not always adequately explained the nature of a watershed management approach. Recently, technical experts have made efforts to explain how watershed management can help in land stewardship efforts that are focused on securing the flows of natural and economic resources, generating investment, employment, and income opportunities, and maintaining environmental quality—all within a sustainable framework. This book is one such effort.
- The downstream benefits of watershed management have been doubted, and unfortunately, little quantitative information concerning the effects is available. One reason for the information deficiency is a lack of monitoring, which, if accomplished, would provide the necessary data to evaluate the effectiveness of watershed management practices, projects, and programs (Doolette and Magrath, 1990).
- Many people do not separate the human-caused effects of land-use practices from effects caused by natural events. Erosion and sedimentation, flooding, and landslides

occur naturally. For example, natural geologic erosion results in unstable slopes, landslides, and high levels of sedimentation in areas of active mountain-building. However, human activities on watershed lands can affect the frequency and severity of any naturally occurring event. Deforestation followed by improper agricultural cultivation or overgrazing of livestock influences the occurrence of flooding and sedimentation. A key decision is whether the potential exists to improve the conditions through changes in land use and the introduction of effective watershed management.

- Unrealistic expectations from watershed management programs have led to suggestions that watershed management does not meet planned goals and objectives. For example, the planners of watershed management practices implemented to reduce accumulations of sediment in downstream reservoirs must consider the natural erosion-sedimentation processes, amount of the watershed area affected by the practices, proximity of the practices to the reservoir, existing levels of sediment in the stream channels, and other land-use practices on the watershed.
- A lack of technical expertise in planning, implementing, and monitoring watershed management programs exists. Participation of professionals experienced in the disciplines associated with watershed management, such as hydrology, geology, soil science, forestry and rangeland management, and agronomy, is needed when the watershed management program goals are being developed. This expertise has not always existed on the teams that have developed past watershed management programs.

6.3.2. Overcoming Barriers

Barriers to the wider adoption of a watershed management approach to land stewardship are slowly breaking down. Most of the planners, managers, and decision-makers that are concerned about future land stewardship recognize the imperative of environmentally sound sustainable natural resource development and use (Gregersen et al., 1987; Brooks et al., 1992, 1994; Quinn et al., 1996). Ignoring the boundaries and interrelationships established by nature will inevitably lead to serious consequences.

Watershed management is not an attempt or claim to change the world by replacing current land-use practices with a cure-all watershed management formula. Additionally, the development of natural and economic resources within a watershed management framework does not mean only professional watershed managers should direct the planned watershed management practices, projects, and programs. Instead, watershed management is an integral component in development programs that focus on water resources, forestry, agricultural, and related land and resource uses. To be effective, land-use administrators, water resource managers, foresters, and agriculturalists, along with professional watershed managers, must sustain these components.

Overcoming the barriers to land stewardship also requires responsible government agencies, locally-led partnerships, councils, and corporations, and other institutions to:

- Increase the awareness of all stakeholders about the importance of sustainable land use and the relationships that watershed management is built on including the biophysical realities and the economic, social, and cultural factors that affect land use on watershed lands.

- Identify all stakeholders, including the upstream and downstream stakeholders, in watershed-use issues, and their perceptions and motivations about the issues.
- Clarify agency or institutional jurisdiction over watershed management activities and improve the coordination between the agencies and institutions. This is especially significant because few countries, including the United States, have one agency or institution charged with natural resource management on a watershed scale. Most countries, have several agencies or institutions that have the jurisdiction over uplands and particular activities in the upland areas.
- Facilitate local management of upland natural resources by local residents on watersheds that are partially or entirely privately owned or controlled, and where agencies are not responsible for land, water, or other natural resource management.
- Fairly distribute the benefits (positive impacts) and costs (negative impacts) associated with upland natural resource use, and the application of watershed management practices between the upland and downstream land users and other stakeholders.
- Assess the short- and long-term impacts of watershed management policies and activities as they evolve to encourage more effective watershed management. Feedback mechanisms (monitoring and evaluation programs) for this assessment must determine whether commodity producing activities and the soil and water resources on which these activities depend can be sustained under the current policies—results from the assessment must be incorporated into future land-use policies.

6.4. SUMMARY

Protocols for adopting a watershed management approach to land stewardship include anticipating future landscape conditions, responding to increased water demands, and implementing effective policies. Anticipating future landscape conditions is a difficult task. Nevertheless, most people agree that landscapes should be managed to recognize and sustain inherent ecological properties, the composition and functioning of the existing ecosystems, and society's perceptions of how the landscapes should be used.

Responding to increased demands for water will continue to be the main focus of future watershed management, largely because water unifies the many facets of integrated land stewardship. Addressing the supply and demand side of water use is also necessary. Implementing land-use policies that incorporate ecological understanding into their framework and promoting democratic ideals on how land, water, and other natural resources should be used is essential. Local commitment and participation within social, political, and institutional settings are important to implement the policies successfully and attain stakeholder compliance. Barriers to the adoption of a watershed management approach to land stewardship are being overcome through the combined efforts of technically-trained planners and manager, decision-makers, locally-led advocacy groups, and other stakeholders.

REFERENCES

Brooks, D. J., Grant, G. E., *New Perspectives in Forest Management: Background, Science Issues, and Research Agenda*. Portland, OR: Pacific Northwest Forest and Range Experiment Station, USDA Forest Service, Research Paper PNW-RP-456, 1992.

Brooks, K. N., Ffolliott, P. F., Gregersen, H. M., DeBano, L. F., *Hydrology and the Management of Watersheds*. Ames, IA: Iowa State University Press, 1997.

Brooks, K. N., Gregersen, H. M., Ffolliott, P. F., Tejwani, K. G. "Watershed Management: A Key to Sustainability." In *Managing the Worlds's Forests: Looking for Balance Between Conservation and Development,* N. P. Sharma, ed. Dubuque, IA: Kendall/Hunt Publishing Company, 1992, pp. 455-487.

Brooks, K. N., Ffolliott, P. F., Gregersen, H. M., Easter, K. W., *Policies for Sustainable Development: The Role of Watershed Management*. Washington, DC: U.S. Department of State, EPAT Policy Brief 6, 1994.

Brooks, K. N., Gregersen, H. M., Lundgren, A. L., Quinn, R. M., *Manual on Watershed Management Project Planning, Monitoring and Evaluation*. College, Laguna, Philippines: ASEAN-US Watershed Project, 1990.

Brown, R. S., Marshall, K. Ecosystem management in state governments. Ecological Applications 1996; 6:721-723.

Cortner, H. J., Moote, M. A. "Ensuring the Common for the Goose: Implementing Effective Watershed Policies." In *Land Stewardship in the 21st Century: The Contributions of Watershed Management,* P. F. Ffolliott, M. B. Baker, Jr., C. B. Edminster, M. C. Dillon, and K. L. Mora, tech. coords. Fort Collins, CO: Rocky Mountain Research Station, USDA Forest Service, Proceedings RMRS-P-13, 2000, pp. 247-256.

de Steiguer, J. E. "Sustaining Flows of Crucial Watershed Resources." In *Land Stewardship in the 21st Century: The Contributions of Watershed Management,* P. F. Ffolliott, M. B. Baker, Jr., C. B. Edminster, M. C. Dillon, and K. L. Mora, tech. coords. Fort Collins, CO: Rocky Mountain Research Station, USDA Forest Service, Proceedings RMRS-P-13, 2000, pp. 215-220.

Doolette, J. B., Magrath, W. B., *Watershed Development in Asia: Strategies and Technologies*. Washington, DC: The World Bank, World Bank Technical Paper 127, 1990.

Easter, K. W., Hufschmidt, M. M., *Integrated Watershed Management Research for Developing Countries*. Honolulu, HA: Environment and Policy Institute, East-West Center, 1985.

Easter, K. W., Cortner, H. J., Seasholes, K., Woodard, G., *Water Resources Policy Issues: Selecting Appropriate Options*. St. Paul, MN: University of Minnesota, EPAT/MUCIA Draft Policy Brief, 1995.

Ffolliott, P. F., DeBano, L. F., Ortega-Rubio, A. "Relationship of Research to Management in the Madrean Archipelago Region." In *Biodiversity and Management of the Madrean Archipelago: The Sky Islands of Southwestern United States and Northwestern Mexico*, L. F. DeBano, Ffolliott, P. F., Ortega-Rubio, A., Gottfried, G. J., Hamre, R. H., Edminster, C. B., tech. coords. Fort Collins, CO: Rocky Mountain Forest and Range Experiment Station, USDA Forest Service, General Technical Report RM-GTR-264, 1995, pp. 31-35.

Ffolliott, P. F., Baker, M. B., Jr., Lopes, V. L. "Watershed Management Perspective in the Southwest: Past, Present, and Future." In *Land Stewardship in the 21st Century: The Contributions of Watershed Management,* P. F. Ffolliott, M. B. Baker, Jr., C. B. Edminster, M. C. Dillon, and K. L. Mora, tech. coords. Fort Collins, CO: Rocky Mountain Research Station, USDA Forest Service, Proceedings RMRS-P-13, 2000b, pp. 30-36.

Ffolliott, P. F., Baker, M. B., Jr., Edminster, C. B., Dillon, M. C., Mora, K. L. Mora, tech. coords., *Land Stewardship in the 21st Century: The Contributions of Watershed Management*. Fort Collins, CO: Rocky Mountain Research Station, USDA Forest Service, Proceedings RMRS-P-13, 2000a.

Gottfried, G. J., Eskew, L. G., Curtin, C. G., Edminster, C. B., compilers, *Toward Integrated Research, Land Management, and Ecosystem Protection in the Malpai Borderlands: Conference Summary*. Fort Collins, CO: Rocky Mountain Research Station, USDA Forest Service, Proceedings RMRS-P-10, 1999.

Gottfried, G. J., Edminster, C. B., Bemis, R. J., Allen, L. S., Curtin, C. G. "Research Support for Land Management in the Southwestern Borderlands." In *Land Stewardship in the 21st Century: The Contributions of Watershed Management,* P. F. Ffolliott, M. B. Baker, Jr., C. B. Edminster, M. C. Dillon, and K. L. Mora, tech. coords. Fort Collins, CO: Rocky Mountain Research Station, USDA Forest Service, Proceedings RMRS-P-13, 2000, pp. 330-334.

Gregersen, H. M., Easter, W. K., de Steiguer, J. E. "Responding to Increased Needs and Demands for Water." In *Land Stewardship in the 21st Century: The Contributions of Watershed Management,* P. F. Ffolliott, M. B. Baker, Jr., C. B. Edminster, M. C. Dillon, and K. L. Mora, tech. coords. Fort Collins, CO: Rocky Mountain Research Station, USDA Forest Service, Proceedings RMRS-P-13, 2000, pp. 238-246.

Gregersen, H. M., Brooks, K. N., Dixon, J. A., Hamilton, L. S., *Guidelines for Economic Appraisal of Watershed Management Projects*. Rome: Food and Agriculture Organization of the United Nations, FAO Conservation Guide 16, 1987.

Gregersen, H. M., Brooks, K. N., Ffolliott, P. F., Lundgren, A. L., Easter, K. W., Belcher, B., Eckman, K., Quinn, R., Ward, D., White, T. A., Josiah, S., Xu, Z., Robinson, D., *Assessing Natural Resources Policy Issues: A Framework for Developing Options*. St. Paul, MN: University of Minnesota, EPAT/MUCIA Draft Policy Brief, 1994.

Hernandez, M., Miller, S. N. "Integrated Landscape/Hydrologic Modeling Tool for Semiarid Watersheds." In *Land Stewardship in the 21st Century: The Contributions of Watershed Management,* P. F. Ffolliott, M. B. Baker, Jr., C. B. Edminster, M. C. Dillon, and K. L. Mora, tech. coords. Fort Collins, CO: Rocky Mountain Research Station, USDA Forest Service, Proceedings RMRS-P-13, 2000, pp. 320-324.

Hydrology and Water Resources Programme of the World Meteorological Organization, *The Dublin Statement on Water and Sustainable Development*. http://www.wmo.ch/web/homs/icwedece.html#introduction, 1999.

Irland, L. C. Getting from here to there: Implementing ecosystem management on the ground. Journal of Forestry 1994; 92(8):12-17.

Jones, J. R., Martin, R., Bartlett, E. T. Ecosystem management: The U.S. Forest Service's response to social conflict. Society and Natural Resources 1995; 8:161-168.

Kessler, W. B., Salwasser, H., Cartwright, C. W., Caplan, J. A. New perspectives for sustainable natural resources management. Ecological Applications, 1992; 2:221-225.

Lant, C. L. Introduction: Human dimensions of watershed management. Journal of the American Water Resources Association 1999; 35:483-486.

McDonald, B. "The Formation and History of the Malpai Borderlands Group." In *Biodiversity and Management of the Madrean Archipelago: The Sky Islands of Southwestern United States and Northwestern Mexico*, L. F. DeBano, Ffolliott, P. F., Ortega-Rubio, A., Gottfried, G. J., Hamre, R. H., Edminster, C. B., tech. coords. Fort Collins, CO: Rocky Mountain Forest and Range Experiment Station, USDA Forest Service, General Technical Report RM-GTR-264, 1995, pp. 483-486.

McDonald, B. "Anticipating Future Landscape Conditions: A Case Study." In *Land Stewardship in the 21st Century: The Contributions of Watershed Management,* P. F. Ffolliott, M. B. Baker, Jr., C. B. Edminster, M. C. Dillon, and K. L. Mora, tech. coords. Fort Collins, CO: Rocky Mountain Research Station, USDA Forest Service, Proceedings RMRS-P-13, 2000, pp. 235-237.

Miadenoff, D. J., Pastor, J. "Sustaining Forest Ecosystems in the Northern Hardwood and Conifer Region: Concepts and Management." In *Defining Sustainable Forestry*, G. H. Aplet, Johnson, N., Olson, T., Sample, V. A., eds. Washington, DC: Island Press, 1993, pp. 145-180.

National Research Council, *Forestry Research: A Mandate for Change*. Washington, DC: National Academy of Science Press, 1990.

Quinn, R. M., Brooks, K. N., Ffolliott, P. F., Gregersen, H. M., Lundgren, A. L., *Reducing Resource Degradation: Designing Policy for Effective Watershed Management*. Washington, DC: U.S. Department of State, EPAT Working Paper 22, 1995.

Ridd, M. K., *Area-Oriented Multiple Use Analysis*. Ogden, UT: Intermountain Forest and Range Experiment Station, USDA Forest Service, Research Paper INT-21, 1965.

Salwasser, H., MacCleery, D. W., Snellgrove, T. A. "An Ecosystem Perspective on Sustainable Forestry and New Directions for the U.S. National Forest System." In *Defining Sustainable Forestry*, G. H. Aplet, Johnson, N., Olson, T., Sample, V. A., eds. Washington, DC: Island Press, 1993, pp. 44-89.

Staebler, R. N. Ecosystem management: An evolving process. Journal of Forestry 1994; 92(8):5.

Swank, W., VanLear, D., eds. Multiple-use management: Ecosystem perspectives of multiple-use management. Ecological Applications 1992; 3:219-274.

The World Bank, *Water Resources Management*. Washington, DC: The World Bank, 1993.

Watershed Group on Watershed Management and Development, *The Role of Watershed Management in Sustainable Development*. St. Paul, MN: Forestry for Sustainable Development Program, University of Minnesota, Working Paper 3, 1988.

Webster, H. H., Chappelle, D. E. The curious state of forestry in the United States. Renewable Resources Journal 1997; 15(1):6-8.

Wilhite, D. A. State actions to mitigate drought: Lessons learned. Journal of the American Watershed Resources Association 1997; 33:961-967.

INDEX